MAYA
DAYKEEPING

MAYA

Three Calendars from Highland Guatemala

DAYKEEPING

John M. Weeks, Frauke Sachse, and Christian M. Prager

UNIVERSITY PRESS OF COLORADO

Published by the University Press of Colorado
5589 Arapahoe Avenue, Suite 206C
Boulder, Colorado 80303

 The University Press of Colorado is a proud member of
the Association of American University Presses.

The University Press of Colorado is a cooperative publishing enterprise supported, in part, by Adams
State College, Colorado State University, Fort Lewis College, Mesa State College, Metropolitan State
College of Denver, University of Colorado, University of Northern Colorado, and Western State
College of Colorado.

∞ This paper meets the requirements of the ANSI/NISO Z39.48-1992 (Permanence of Paper).

Library of Congress Cataloging-in-Publication Data

Maya daykeeping : three calendars from highland Guatemala / John M. Weeks, Frauke Sachse, and
Christian M. Prager.
    p. cm. — (Mesoamerican worlds)
Includes bibliographical references and index.
ISBN 978-0-87081-933-9 (hardcover : alk. paper) — ISBN 978-1-60732-246-7 (pbk. : alk. paper)
1. Maya calendar. 2. Maya philosophy. 3. Maya mythology. 4. Mayan languages—Writing. I. Weeks,
John M. II. Sachse, Frauke. III. Prager, Christian M.

F1435.3.C14M39 2009
529'.3297842707281—dc22

                                                                                    2008052158

Design by Daniel Pratt

CONTENTS

FIGURES

TABLES

PREFACE

Large numbers of heretofore unpublished manuscripts in indigenous languages of southern Mexico and northern Central America are held in libraries and other repositories outside of the region known as Mesoamerica. In the United States, for example, sizeable manuscript collections are held at Harvard University, Indiana University, Princeton University, Tulane University, and University of Pennsylvania, among others. Taken as a whole, this material has largely been ignored, mostly because scholars have been unaware of its existence. This volume seeks to publish Mayan transcriptions, accompanied by English language translations, for three divinatory calendars held at the library of the University of Pennsylvania Museum of Archaeology and Anthropology. We hope that we can continue to issue transcripts or facsimiles of the many extraordinary manuscripts presently curated at the Museum Library.

This book began as a modest project to publish a description and analysis of a single K'iche' Maya calendar currently in the collections of the University of

Pennsylvania Museum Library. The manuscript, edited by Oliver La Farge, Rudolf Schuller, and J. Alden Mason, is titled "*Cholbal K'ih* and *Ahilabal K'ih*: Anonymous Quiché Manuscripts." The typescript was finished in 1958, complete with printer's instructions. For a variety of reasons, however, the manuscript was never published and since that time has languished in the library. The history of this manuscript is summarized in the introductory essay.

After some preliminary work Weeks learned that Frauke Sachse and Christian M. Prager, both at the Institut für Altamerikanistik and Ethnologie, Friedrich-Wilhelms Universität in Bonn, Germany, were working on a translation of the 1722 K'iche' calendar under the supervision of Prof. Berthold Riese. A decision was made to combine efforts and produce a single volume.

The idea to publish the La Farge, Schuller, and Mason manuscript was abandoned after it was determined that this translation of the 1722 calendrical document, as well as that of Rudolf Schuller upon whose work it was based, had inaccurate translations. Sachse then agreed to prepare new translations for the 1722 K'iche' calendar, as well as two additional manuscripts, a 1685 Kaqchikel and an 1854 K'iche' calendar.

Individual contributions to the volume varied. Weeks prepared the introductory essay to which Sachse made some additions. Prager provided most of the descriptive narrative presented for the 1722 K'iche' calendar in the introductory essay. Sachse prepared the translations and notes for the 1685 Kaqchikel and 1722 K'iche' calendars. Sachse and Weeks collaborated on the translation of the 1854 K'iche' calendar. Prager drew the regional map, and Sachse prepared the images of the 1854 calendar wheels for publication.

We thank Dr. Richard Hodges, Williams Director of the University of Pennsylvania Museum, for his support in making the publication of this project possible. Wendy Ashmore and Garrett Cook read an earlier version of the manuscript and offered many excellent editing suggestions.

MAYA
DAYKEEPING

Three K'iche'an Divinatory Calendars

The three divinatory calendars presented in this volume are examples of a K'iche'an[1] literary tradition that includes the *Popol Vuh*, *Annals of the Cakchiquels (Memorial de Solola)*, and the *Titles of the Lords of Totonicapan*. Two of the calendars were written in indigenous Kaqchikel or K'iche' languages, but in European script, sometime before or during the eighteenth century. The third example was written in K'iche' and Spanish in 1854. They demonstrate that although linguistic and literary traditions were still being adhered to, there was at the same time an obvious element of adaptation and acculturation, the use of European script.

Calendars such as these continue to be the basis for prognostication or determining the favorable or unfavorable nature of specific periods of time. According to the favor of the days, land may be purchased, sales made in the market, profit accrued, and other economic enterprises pursued. The calendar designates the time for planting and harvest and other agricultural pursuits. The disposition of the days can maintain health and foretell illness or death, influence the naming of children, guide betrothal and marriage. Obligations to the dead are fulfilled on days affiliated with the souls of the ancestors.

These little-known works appear to have escaped the notice of most scholars. Except for occasional mention of their existence, and an unpublished study of the 1722 calendar by Rudolf Schuller and Oliver La Farge (1934), no further work has been done.[2] Although they languished in the library of the University of Pennsylvania Museum for over a century, these are important documents, shedding light on seventeenth-, eighteenth-, and nineteenth-century divinatory practices, and can serve as a basis of comparison with other sources on which our knowledge of K'iche'an divination is based.

MESOAMERICAN DIVINATORY CALENDARS

The indigenous peoples of Mesoamerica in ancient times and in many places into the present, maintained intricate calendars consisting of civil or solar and of sacred or divinatory cycles. The calendar was a foundational achievement of Mesoamerican civilization, reaching its highest elaboration among the Maya of the Classic period. From the earliest times the Maya observed and measured various natural cycles, particularly those related to the astronomical movements of the sun, moon, Venus, and other celestial bodies. The study of the movement of various celestial bodies produced several time cycles.

The civil calendar was a 365-day solar calendar containing eighteen months of twenty days with five days remaining. Each year was given the name of the day which started it, there being only four of the twenty that could appear as the first day of the new year (Table 1.1). These four days—No'j, Iq', Kej, and E—were repeated until after thirteen years the number 13 was reached, at which time the next year began with number 1 again.

The sacred divinatory calendar was not marked off into months but was a combination of day designations created by the coincidence of a number from 1 to 13 with one of the twenty possible names (Table 1.2). This process created different combinations of numbers and names, which were repeated indefinitely to form a cycle of 260 (13 × 20) different days. This period is referred to by scholars as a *tzolk'in*, although in K'iche'an languages it is known as *chol q'ij*, a term meaning the "order of the days," since it serves to designate a series of 260 days not repeated until the beginning of another series of similar duration and having the same numbers and names as the first.

The two cycles, one of 365 days and the other of 260 days, meshed to produce a calendar round to form a period of 18,980 days.

Important dates or period endings in all these calendars were used by the Classic period Maya to commemorate significant events in the lives of important people, such as births, deaths, successions to office, and sacrifices or other rituals.

Table 1.1. Day names in the 1685, 1722, and 1854 divinatory calendars.

1685 Kaqchikel	1722 K'iche'	1854 K'iche'
Ymox	Ymox / Ymos	Ymux
Yɛ	Yɛ	Yc
Aɛbal	Aɛbal / Akbal / Acbal	Bacbal
Kat	4at	Cat
Can	Can	Kan
Camey	Queme / Came	Kame
Quieh	Queh	Quiej
Kanel	ɛanil / Canil	Kanil
Toh	Toh / Thoh	Toj
Tzij	4'ij	Tzii
Batz	Ba4,	Batz
Ee	Ee / E	Ec
Ah	Ah	Ah
Yix	Yx / Balam	Yx
Tz'iquin	4,iquin	Tziquin
Ahmak	Ahmak	Ahmac
Noh	Noh	Noh
Tihax	Tihax	Tihax
Caok	Caok	Caquok
Hunahpu	Hunahpu	Ahpu

Some activities appear to have been timed to correlate with specific cycles; for example, some war events are associated with the cycle of the planet Venus.

An exclusively Classic period Maya calendrical achievement was the long count, which permitted an infinite computation of time from an established mythical starting point, backward or forward. The long count is a linear count of days that began in 3114 BC.

Given the ancient importance of the calendar, one might wonder why the indigenous calendar did not persist more strongly after the Spanish conquest in the sixteenth century than it did. Much of calendrical knowledge was probably held by a small group of individuals who guarded that knowledge but were easily singled out for control, suppression, or elimination. Calendrical knowledge was a prime source of socioreligious power and was, along with the practice of human sacrifice, a major target of early Spanish missionaries, who quickly substituted saints' days and other Catholic ritual occasions for indigenous ceremonies. The religious brotherhood dedicated to the cult of a specific saint (*cofradia*) and the Gregorian calendar were the chief instruments for effecting these changes. Some small-scale rituals for crops and households survived. In parts of the K'iche'an area, the ritual calendar has persisted and is still used in these smaller-scale rituals.

La Farge (1947:180–181) has noted similarities in divinations between the Codex Dresden and the Ajilab'al q'ij from the 1722 K'iche' calendar. Codex

Table 1.2. Month names in the 1685 and 1722 divinatory calendars.

1685 Kaqchikel	1722 K'iche' A-I	1722 K'iche' A-II	1722 K'iche' A-III-VI	1722 K'iche' C
Nabeimam	Nabemam	Nabe Má	Nabe Mam	Mam
Rucabmam	Vcab mam	Vcab Má	Vcab Mam	Vcab Mam
Liεinεa	Liqinca	Liquinca	Nabe Liquin Ca	Liquin Ca
Nabeitoεiε	Vcab Liquinca	Vcab Liquinca	Vcab Liquin Ca	Vcab Liquin Ca
Rucatoεiε	Nab Pach	Nabe Pach	Nabe Pach	Pach
Nabeipach	Vcab Pach	Ucab Pach	Vcab Pach	Vcab Pach
Rucanpach	4,içilakam	4,içilakam	4isi Lakam	4,içilakan
4,iquin εih	4,iquin εih	4,iquin εih	4,iquin εih	4,iquin εih
Cakan	Cakam	Cakam	Cakam	Cakam
Ibota	Botam	Botam	Botam	Botam
Katic	Nabe εih	Nabeçih	Nabeçih	Çih
Yzcal	Ucab Çih	Vcabçih	Vcabeçih	Vcab Çih
	Rox Çih	Roxçih	Rox Çih	Urox Çih
Pariche	Chee	Chee	Chee	Chee
Tacaxepual	Tequexepual	Tequexe pual	Tequexe pual	Tequexpual
Nabeitumuzuz	4,ib'apopp	4,ibapp	4ibapopp	4,ibapop
Rukantumuzuz	Cac	Çak	Çak	Çak
Cibixiε	4hab	4hab	4hab	4hab
Vchum				
Tzapiεih	4isbal rech	4,apiεihih	4,apiεih	4,api εih

Dresden, one of the four surviving prehispanic Maya codices, consists of thirty-nine leaves painted, in color, on both sides with glyphs and portraits of deities. The contents are divided into several major parts, including seventy-six 260-day almanacs and 364-day counts of divination that indicate good, bad, and indifferent days and the benevolent or adverse influence of the presiding deities in matters of agriculture, weather, disease, and medicine. The codex is a condensed book of divination of good and bad days for human enterprise with directions to propitiate the gods (Thompson 1972). Similarly, the Books of Chilam Balam, written in the Yucatec Maya language anywhere from the sixteenth to the nineteenth centuries, are a genre unique to post-conquest Yucatán. They have been named for the prehispanic Chilam Balam, the Jaguar Prophet, who made prophesies based on historical knowledge and a cyclical view of time. The manuscripts are compilations of history, myth, prognostication, farmers' almanacs, medical diagnoses, and herbal recipes. Each manuscript appears to be a compilation of passages copied from other texts. A great deal of calendrical material, including weather predictions, prognostics of "good" and "bad" days, warnings of sickness and death, and various other portents, occur in various of these Books of Chilam Balam (Scholes et al. 1946). For example, the Chilam Balam of Ixil, which dates from the late eighteenth century, includes a Catholic calendar, which is not translated, giving the days of each month together with the pacts and dominical letters. Except for a few church festivals, the saint for each day is named, which was useful in naming children. Accompanying this calendar is a Maya treatise on the European zodiac. Beneath a picture of each sign is the usual information found in European almanacs, such as the day when the sun enters the sign, the number of stars in that sign, the hours of daylight and darkness, and other information for the guidance of a person born under this sign. To this are added a chart with some tables and other material on zodiacal anatomy, so that a healer might avoid bleeding any part of the body while the sun is passing through the sign of the zodiac ascribed to that part. There are two calendar wheels, one of which represents an alleged *katun*, or period of thirteen years. Similarly, the so-called Codex Perez consists of extracts that Juan Pío Pérez copied from various Maya manuscripts during the second quarter of the nineteenth century. A large part of it was taken from the Book of Chilam Balam of Mani. The first third of the manuscript is composed of Maya translations of European astrological and calendrical material. Much attention is given to augural or divinatory aspects of the Maya calendar.

After the imposition of Spanish rule in the sixteenth century, the calendrical system persisted throughout the Colonial period. In some areas calendrical knowledge was maintained on an oral basis, whereas in others it was retained with the aid of written schematic drawings or calendar wheels. A calendar wheel is a Colonial period image that displays cycles of time in a circular format.[3] Its use is mentioned in the *Annals of the Cakchiquels* (1953:98–159) and other early narratives. The seventeenth- and eighteenth-century historians describe its use (Ximénez 1929–1931, 1:102–103). Pedro Cortés y Larraz (1958:2:57), bishop of Guatemala between 1768 and 1781, undertook an administrative visit to 113 curatos in his dioceses and stated that the traditional calendar was in use "in all the parishes of the K'iche' and Kaqchikel." He made specific reference to the ancient calendar in his descriptions of the parishes of Nuestra Señora de la Concepción de Zamayac, Quezaltenango, San Pedro La Laguna, and Santa Cruz del Quiche.

We find evidence for the retention of the ancient Maya calendar in many contemporary communities (Miles 1952), and these data have great potential for inferences about the function and meaning of the ancient Maya calendar. Scholars have traditionally assumed that in ancient times the common Maya knew little of calendrical ritual. This ignorance was thought to extend to all parts of the calendar, explaining its apparent total disappearance since the conquest. However, evidence collected at least in the highlands of Guatemala by Robert Burkitt (1930–1931), Samuel K. Lothrop (1929, 1930), and Oliver La Farge (1947:75) indicates that the basic components of the calendar were common knowledge. Lothrop reports that the calendar was so vigorously in use that a storekeeper wrote the indigenous dates on his calendar for reference in dealing with the K'iche' Maya of Momostenango (La Farge 1947:75–76). Burkitt (n.d.:387) gives the following assessment of the retention of the indigenous calendar in the Ixil-speaking community of Chajul: "Éh, Noh, Iq, and Txéh, are the lucky days. The other 16 indifferent. Everybody knows these days. Servant girls are hired by periods of 20 days. Certain people make it a business to keep the right count of the days. Today (1913 August 30, Saturday) is Txéh."

Although the calendar is no longer in use in Zinacantán in the southern Mexican state of Chiapas, evidence indicates that the solar-year calendar was in use there as late as 1688; there are no data on the divinatory calendar (Vogt 1969:603–604). Sacristans in the church at Zinacantán are usually literate, and their long experience makes them important advisors (*mayordomos*) to the ranked members of the religious fraternities devoted to the worship of specific saints. Because they are able to read the church calendar, which is printed in Spanish, they are still responsible for telling the mayordomos the dates on which they must perform rituals (Cancian 1965:45). Alfonso Villa Rojas obtained information in 1936–1937 on the nineteen-month calendar and its use in connection with the agricultural round in Oxchuc, a Tzeltal community in Chiapas (Andrade et al. 1938).

Manuel García Elgueta (1962) describes the use of the traditional calendar among the K'iche' during the late nineteenth century. By the middle of the twentieth century the original calendrical system, or parts of it, existed in many communities in the Mexican states of Veracruz, Oaxaca, and Chiapas, as well as the highland region of Guatemala.

Ethnohistorian Suzanna W. Miles (1952:273–275) identified almost ninety communities known to have retained calendars into the middle of the twentieth century. Of these, eighty-two have been described in some detail. In the western highlands of Guatemala a total of thirty-four Chuj, Ixil, Jakaltec (Popti), Mam, and Poqomchi' communities have calendars defined by year bearers, the 365-day year with eighteen cycles of twenty named days, and the thirteen numbers. More to the east in the central highlands of Guatemala are twenty-three Mam, Awakateko, K'iche' and Kaqchikel, and Poqomchi' and Q'eqchi' communities with calendars defined by the 260-day count, the permutation of the twenty named days and the thirteen numbers.

Many other anthropologists have noted the importance of the calendar for contemporary indigenous communities (Bunzel 1952; Burkitt 1930–1931; Falla 1975; Gates 1932a, 1932b; Goubaud Carrera 1935; La Farge 1930, 1947:123, 164; La Farge and Byers 1931:116, 659, 660; Lehmann 1910; Lincoln 1942:103; 1945:121; Lothrop 1930; Miles 1952, 1957, 1965; Rodríguez and Crespo 1957; Rosales 1949a:48, 55; 1949b:683; Sapper 1925; Schultze Jena 1933, 1946; Tax 1947a:34; 1947b:416; Tedlock 1982, 1992; Termer 1930; Thompson 1932; Wagley 1941).

Miles also observed that in areas of highland Guatemala where the thirteen numbers of the 260-day count had been lost, the twenty named days survived as a cycle and assumed the divinatory functions of the 260-day count as a whole. This element, the twenty day names of the 260-day count, is the lowest surviving form of the Maya calendar count and represents the ultimate reduction of the calendrical structure.

Such is also the case with the prognostication tables, which have come down to us in the literary tradition of the Maya of Yucatan. In the Books of Chilam Balam of the eighteenth century we find lists of days, each day with its specific properties and prognostications annotated. These prognostication tables are written in Yukatek by means of an adapted Latin alphabet, but as a comparison with passages of similar content in the Codex Dresden shows, they no doubt have their origin in the hieroglyphic books from prehispanic times (Gubler and Bolles 2000:8–9; La Farge 1947:180–181). The most precise, and also the most extensive, divinatory list in the Books of Chilam Balam is List no. 1 from the Book of the Chilam Balam of K'awa, a small village near Chichén Itzá. It consists of the names of the twenty days and the specific properties that these days have in shaping the destinies, qualities, basic behaviors, and future occupations of men and women who were born under their influence.

CALENDRICAL PRACTICE IN HIGHLAND GUATEMALA

For the past thirty years most of the K'iche'an communities in the central and western highlands have been brutally repressed by the national government of Guatemala, resulting in a reduction of the influence of traditional religion (Figure 1.1). However, daykeepers still remain among the surviving Maya. These calendar priests continue to calculate the days and interpret their qualities in order to reveal answers about mental and physical dispositions, the causes of evil or success and failure of events, and consequently the best day for undertaking such essential activities as planting and harvesting or marriage. Several investigators have fortunately published the results of their field investigations of calendrical divination and its effect on local indigenous society.

North American anthropologist Barbara Tedlock, working with the K'iche' of Momostenango, undertook formal training and, together with her husband, was initiated as a calendar diviner in 1976. Her *Time and the Highland Maya* (1982, 1992) focuses on the concepts and the procedures involved in the training of a K'iche' calendar diviner. It not only presents insights into the significance of ceremonial time, location, and meaning, but it also provides a glimpse at the mental processes involved in the minds of both diviner and client during the process of a calendrical divination.

German anthropologist Eike Hinz spent fifteen months of fieldwork between 1980 and 1983 in the Q'anjob'al community of San Juan Ixcoy in the northwestern highlands of Guatemala, during which time he collaborated with a diviner who used the prehispanic 260-day calendar in his consultations. Hinz examined the Q'anjob'al concept of "illness" and analyzed the psychical, psychotherapeutical,

1.1. HIGHLAND REGION OF GUATEMALA.

and socio-therapeutical effects of healing in calendrical divination. The diviner's consultation constitutes a type of psycho-sociotherapy in which he not only interprets the existential problems and preoccupations of clients but also attempts to resolve them. During fieldwork, Hinz was also trained and initiated by a calendar diviner–healer. In his monograph, *Misstrauen führt zum Tod* (1991), Hinz presents twelve complete cases (of a total of fifty recorded) of calendrical divination and healing. He recorded all divinations and ensuing therapeutical dialogues between healer and patient in Q'anjob'al and then transcribed them in both Q'anjob'al and German.

North American anthropologists Benjamin N. Colby and Lore M. Colby, working during the late 1960s and early 1970s with an Ixil daykeeper, published *The Daykeeper: The Life and Discourse of an Ixil Diviner* (1981), a magnificent study that documents the cultural principles organizing the daykeeper's methods of divination and guiding his interpretation of dreams and his cures for the sick. They identify and define cultural patterns underlying the stories he relates and the morals he draws from them. These patterns are used to inform our perception of the daykeeper's experience of life, and the reader gains an understanding of the relation between culture and thought.

Participatory investigations by Tedlock and Hinz and detailed observations made by Colby and Colby and others (Bunzel 1952; La Farge 1947; La Farge and Byers 1931; Lincoln 1945; Sapper 1925; Schultze Jena 1933, 1946, 1947; Termer 1930) on the beliefs and practices associated with the calendar, particularly the role of the daykeeper, explain much about highland Mayan behavior and ethics.

Divination and the management of time have a fundamental role in Maya culture and have been practiced from ancient times through the present. Historical and ethnographic accounts provide information about diviners and other types of non-Catholic religious specialists (Table 1.3). The shaman priest determines the days on which both communal religious ceremonies and cofradia ceremonies are to be held. The prayer sayer requests good providence or assists in effecting cures of sick clients and functions in dawn ceremonies of various kinds. The daykeeper, or calendar priest, uses a divining bundle of tz'ite' seeds to count the days and make diagnoses. A subcategory of diviner includes those who use crystals instead of tz'ite' seeds.

K'ICHE'AN DAYKEEPERS

Only the *ajq'ij*, daykeeper, or calendar priest, knows how to properly interpret the causal relationship between an event and the quality of the day. The quality of a particular day is defined by the combination of the quality associated with the day sign and the quality associated with its number coefficient. Thus, each day of the 260-day count has its individual prognostication. The day gods reveal the underlying

Table 1.3. Religious specialists identified in K'iche'an dictionaries.

K'iche'an gloss	Description
Ab'ay ruq'a	Soothsayer for luck (Solano 1580)
Ajaq'om	Curer (Coto 1983)
Ajawalin q'ij	Soothsayer with luck or omens (Barrera 1745)
Ajilab'al q'ij	Lot caster; he who counts the days, or omens, or the months in ancient times (Vico 1550)
Ajilanel	Soothsayer for luck (Solano 1580)
Aj'itz	Witch, warlock (Coto 1983; Solano 1580)
Ajkun	Curer (Solano 1580)
Ajmalola'	Soothsayer for luck (Solano 1580)
Ajq'ij	Master of the calendar, soothsayer, diviner (Barrera 1745; Coto 1983; Solano 1580; Vico 1550); witch, warlock (Coto 1983; Solano 1580); the K'iche' high priest who keeps the day count and decides on the dates of ceremonies (Vico 1550)
Ajtoq'ol	Soothsayer (Solano 1580; Vico 1550)
B'alam	Witch (Solano 1580)
Cholol q'ij	There are Indians who, according to the count of the days, know and announce (Coto 1983); soothsayer; they are esteemed among the Indians as those who count the days and make predictions (Coto 1983)
Cholol tz'itz'	Soothsayer for luck (Solano 1580)
Jalom	Witch, warlock (Coto 1983:74; Solano 1580)
Ki cholo q'ij	Soothsayer for dreams and omens (Solano 1580)
Kinawalin	Soothsayer for luck or omens (Solano 1580)
Kiq'ijin	Soothsayer for luck or omens (Solano 1580)
Kisaqiwachin	Soothsayer for luck or omens (Solano 1580)
Makajik xuqul	Soothsayer for luck (Solano 1580)
Nawal	Witch, warlock (Coto 1983; Solano 1580)
Nik' wachi' ruwach pa ya'	Soothsayer with water (Solano 1580)
Saq tijax	Witch who travels through the sky like a comet (Coto 1983)
Saqiwachinel	Soothsayer (Barrera 1745; Solano 1580)
Tz'etom pa ya'l	Soothsayer who looks in water (Solano 1580)

cause of a sickness or misfortune to the daykeeper, and he passes this information on to the client. Knowing the variables associated with the day, the calendar priest can interpret dispositions and events. Divination is used to determine the cause and cure of sickness, to interpret dreams, to recover lost or stolen objects, and to determine times and places for ritual. A calendar priest is consulted for advice in all situations of emergency and stress, such as illness, land disputes, lost property, commerce, travel, adultery, death, birth, dreams, omens, or marriage (Colby and Colby 1981:222; Tedlock 1982:153).

The calendar and associated beliefs facilitate communication between people and supernaturals. This communication is possible through several techniques, all well documented throughout most of Mesoamerica. These included sortilege, or the casting of a handful of bean-shaped seeds to divine information about past, present, and future; the interpretation of dreams; the reading of signs and omens; the belief in lucky and unlucky days; and observations of reflections.

Divination continues to be a key feature of contemporary Mesoamerican indigenous and non-indigenous magico-religious practice. Maize sortilege is widespread; dream and vision interpretation continue to occupy an important place in indigenous belief. Other common techniques of performing divination, such as crystal gazing, observing reflections in obsidian, breaking eggs, pulsing, and interpreting patterns of water vapor and incense smoke to determine the cause of disease, do not appear to occur among K'iche'an peoples.

The seventeenth-century "Vocabulario de la lengua cakchiquel" by the Franciscan Tomas de Coto, gives the following entry for *pronostico* (prediction): "The ancient vocabulary gives natajik . . . The Indians among them, who prognosticate, are called cholol q'ij, or ajq'ij. Lab', is an omen or bad prognostic . . . wachik' is the vision which they dream . . . they prognosticate with reflections on water . . . the birds foretell rain when they sing" (Coto 1983:443).

For *sortilege*, Coto (1983:110) gives:

> to draw lots or to throw for luck with kernels of maize, or with some beans, or with small sticks of a tree called tz'ite' . . . tin chol ixim, or tz'ite', which is putting these small grains in order, according to counts or sorcery. And, these who do this are called cholol ixim, or cholol tz'ite' . . . This is for the good or bad days according to their astrology . . . And, therefore, to refer to the ancients who knew these days and sorcery, ajq'ij, or cholol q'ij. They use also, with the toes of the foot to count these maize kernels, and with the toes of the feet are able to make their absurd ideas. And these usually make these ancients liars . . . count by maize kernels, to settle accounts, the count of the ancients: tiwiximaj.

Although supernaturals are usually punishing beings, they are often seen as beneficial and will sometimes send warnings through dreams and protect or reward people. Dreams are the private medium to receive instructions and messages from the supernatural power or from one's ancestors or if danger threatens one's family. Various members may dream about the possible causes and seek help accordingly. Although harmless or benevolent dreams are often interpreted by the dreamer himself or an older relative or friend, more complex or threatening dreams are formally presented to the professional interpreter of dreams, the daykeeper. He will take the proper measures needed to restore the personal equilibrium of his clients by searching or reestablishing vulnerable ties in the dreamer's human and superhuman *Umwelt*. Dreams provide a kind of interface between individual concerns and socially shared religious activities.

For the interpretation of signs and omens, Coto (1983:522) writes,

According to the astrology of these Indians, there are many falsehoods given when they are counting the days or predicting because each town has its own way of counting. According to the predictions based on the signs of their calendar, they use the name of the day of birth for the people who were born on that day. All of the days of the year have a sign. The boys receive the name of the sign without adding anything; for the girls, they add an x before the name; for example, for a boy they give Pedro Kanel; for a girl, Maria Xkanel . . . B'alam, Kan, K'at, Kamey, Q'anil, Iq' are the names of the days. Likinka, is the name of a month in which they sow their maize; Mam, is a bad sign, because those who are born under this day stay deteriorated; Uchum is a day good for storage and to sow vegetables; Tumuxux, this is for the first rains of the winter; Moh, Pay, and Pach are other month names; Taxepual, is also for the sowing of milpas; Tz'api Q'ij, is the name for the five missing days; Tz'ikin Q'ij, is another month name. These are the principal ones, or most of them. With the count they make the first K'at or B'alam, the second and third, and then they adjust the days of the year. And some last 20 days and some others, seven, which we, including me, never understood, although they have their masters in this faculty who are called Cholol Q'ij, those who announce the days. And among them the names of these soothsayers are kept secret, and are never revealed. They are held in reverence, and are consulted about illnesses and future contingencies, in which they intervene with some deceit of the Devil. The sign of a birth is called, generally, Ruq'ij wi, or Ralaxik wi, Rik'il, Ru chumilal P[edr]o, Juan, etc.

Births and occurrences of general interest were recorded and even predicted using the calendar. Both the K'iche' and Kaqchikel gave to their children the name of the calendar day on which they were born. "The personal name was always that of the day of birth, this being adopted for astrological reasons. There was a perception that the temperament and fortunes of the individual were controlled by the supposed character of his birthday, and its name and number were therefore prefixed to his family name" (Coto 1983:372). "They have surnames which they take from the day or sign of their birth, which are many, according to the count of their calendar; in the sign they are able to find meaning. They also have the surnames of their chinamitales or parcialidades which are signs as well" (Coto 1983:482).

PROGNOSTICATION

The twenty day gods occupy an important place in K'iche'an religion. All ritual, whether for the family, neighborhood groups, or the entire municipality, is done according to the calendar. The twenty day gods have differing attributes and domains of operation. Each god is said to reign when his day comes, and attributes and special interests of the god are associated with the day. The following compilation presents information from several sources concerning the characters of the various day gods. "Good" days are for ceremonies of commemoration; "bad" days are for ceremonies of defense.

The numbers associated with day names also have meaning. According to Bunzel (1952:283), the low numbers of 1, 2, and 3 are "gentle," that is, days for giving thanks and asking for favors. The high numbers of 11, 12, and 13 are "violent," and these are days for "strong ceremonies" involving defense, vengeance, and evil sorcery. The middle numbers of 7, 8, and 9 are "indifferent," neither gentle nor violent. It is on these middle days that regularly recurring ritual to ensure tranquility of life is performed. These are rituals of "commemoration" rather than rituals of "personal crisis."

- *Junapu* or *Ajpu'*, indifferent day (Hernández Spina 1854). Symbolic of the punitive power of the ancestors, embodied in their ownership of house and hearth (Bunzel 1952:280; Edmonson 1997:121; Schultze Jena 1946:37; Tedlock 1992:124–25).

- *Imox*, bad day; the priests of the sun, the Ajq'ijab', on this day pray to the spirits of evil against their enemies (Hernández Spina 1854). A bad and dangerous day, symbolic of the hidden forces in the universe made manifest in insanity. Divinations in 7 Junapu, 8 Imox, and 9 Iq' indicate a failure or confusion regarding the idols in one's house (Bunzel 1952:280; Edmonson 1997:121; Schultze Jena 1946:34; Tedlock 1992:125–126).

- *Iq'*, bad day, the same as preceding (Hernández Spina 1854). Bad day, symbolic of the destructive forces of the universe embodied in the stone idols, and they must be honored with offerings of incense, aguardiente, roses, pine needles, and candles. Painful swellings and cancer are attributed to this day (Bunzel 1952:280–281; Edmonson 1997:121; Schultze Jena 1946:35; Tedlock 1992:126–127).

- *Aq'ab'al*, bad day; the Ajq'ijab' seek the shrines against their enemies (Hernández Spina 1854). Symbolic of evil, the day of slanderers. The day 8 Aq'ab'al is a time to ask protection against slanderers. The days 12 and 13 Aq'ab'al are strong days for working evil against others, and for requesting justice that enemies be punished (Bunzel 1952:281; Edmonson 1997:121; Schultze Jena 1946:35; Tedlock 1992:108–110).

- *K'at*, bad day, the same as the preceding (Hernández Spina 1854). Symbolic of evil in general. The days 7 Aq'ab'al and 8 K'at were bad days when one can pray for protection against the envy of others (Bunzel 1952:281; Edmonson 1997:121; Schultze Jena 1946:35; Tedlock 1992:110–111).

- *Kan*, bad day, the same as the preceding (Hernández Spina 1854). Bad day that brings sickness and is symbolic of the arbitrary cruelness of the universe (Bunzel 1952:281; Edmonson 1997:121; Schultze Jena 1946:35; Tedlock 1992:111–112).

- *Kame* or *Keme*, bad day, the same as the preceding (Hernández Spina 1854). Symbolic of the ultimate dissolution of all things, good and evil, in death. The day above all others for forgiveness or pardon for all the evil deeds that one has committed. Bunzel notes that for some it is a good day, for it signifies that one's evil deeds will be forgiven and that sickness will pass (Bunzel 1952:281–282; Edmonson 1997:121; Schultze Jena 1946:35; Tedlock 1992:112–113).

- *Keej* or *Kiej*, good day, on which beneficial things are asked for the supplicant (Hernández Spina 1854). A good day above all others for requesting favors, it symbolizes the transfiguration and fulfillment in death, as manifested in the ancestors. The day 8 Kej is the day of commemoration of the ancestors (Bunzel 1952:282; Edmonson 1997:121; Schultze Jena 1946:35; Tedlock 1992:113–114).

- *Q'anil*, good day, sacred to the spirits of agriculture; on this day are supplicated all those things that serve man's sustenance (Hernández Spina 1854). A good day symbolic of the regeneration of the earth, of rebirth after death, and seen in the growth of corn. It is the day of the milpa. After the harvest one waits for the day Q'anil to give thanks (Bunzel 1952:282; Edmonson 1997:121; Schultze Jena 1946:35; Tedlock 1992:114–115).

- *Toj*, bad day; unfortunate he who is born thereon; by inevitable destiny he is doomed to be perverse (Hernández Spina 1854). A bad day, a day of sickness, symbolizes the suffering caused by sin. Toj is also the day for calling sickness to punish an enemy (Bunzel 1952:282; Edmonson 1997:121; Schultze Jena 1946:35–36; Tedlock 1992:115–116).

- *Tz'i'*, bad day; on it is sought the undoing of one's enemies (Hernández Spina 1854). An evil day symbolizes sin, especially sexual impurity. There are no ceremonies on this day because it is evil (Bunzel 1952:283; Edmonson 1997:121; Schultze Jena 1946:36; Tedlock 1992:116).

- *B'atz'*, bad day, on which sicknesses, and particularly paralysis, is prayed to fall on one's enemies (Hernández Spina 1854); good day, symbolic of continuity with the past or ancestors (Bunzel 1952:277–278; Edmonson 1997:121; Schultze Jena 1946:36; Tedlock 1992:116–117).

- *E*, good day; on this day contracts of marriage are entered into, preceded by many sacrifices to the benign powers (Hernández Spina 1854); good day, symbolic of destiny as embodied in the day of birth, of one's personality and fortune (Bunzel 1952:278; Edmonson 1997:121; Schultze Jena 1946:36; Tedlock 1992:117–118).

- *Aj*, again a good day, and also consecrated to the gods of agriculture, and to those presiding over the flocks and domestic animals (Hernández Spina 1854). Symbolic of destiny as embodied in the nawal; the day of one's destiny (Bunzel 1952:278; Edmonson 1997:121; Schultze Jena 1946:36; Tedlock 1992:118–119).

- *I'x*, good day. This day is sacred to the spirits of the mountains and forests; on it protection is sought for flocks and animals at the favor of those spirits who rule over the wolves and other carnivorous beasts (Hernández Spina 1854). Symbolic of the creative forces of the universe as embodied in the concept of the earth. The days 8 I'x and 9 Tz'ikin are days when people give thanks for their lodging, for land acquired either by inheritance or by purchase, and to the former owners of the land (Bunzel 1952:279; Edmonson 1997:121; Schultze Jena 1946:36; Tedlock 1992:119–120).

- *Tz'ikin*, most excellent of days. On this day double offerings are made; in the church of the good and supreme deity and to the saints in the churches; also offerings are made in the caves, the profound barrancas, and in deep and somber woods. On this day they pray for all that is beneficial and useful to man; also seek pardon for all sins against the two great powers, the Good and the Evil. This is also the day for the conclusion of marriage contracts and for the beginning of all important affairs (Hernández Spina 1854). Symbolic of good luck in material affairs, including money. The days 8 I'x and 9 Tz'ikin are days to give thanks for one's lodging; the days 7 I'x and 8 Tz'ikin are days to give thanks and to ask for good fortune in money (Bunzel 1952:279; Edmonson 1997:121; Schultze Jena 1946:36; Tedlock 1992:120–121).

- *Ajmaq*, also a most excellent day, the same as the one before. It is also especially consecrated to the spirits presiding over good health (Hernández Spina 1854). A day without defined character but symbolic of the moral forces as embodied in penitential rituals. The days 7 Tz'ikin, 8 Ajmaq, and 9 No'j are days in which to pray for protection, for from these days evil may come to one for his sins or his evil thoughts (Bunzel 1952:279; Edmonson 1997:121; Schultze Jena 1946:36; Tedlock 1992:121–122).

- *No'j*, a propitious day, dedicated to the presiding genius of the soul. On this day they pray that the suppliant and his family may be endowed with good judgment (Hernández Spina 1854). This day is symbolic of the ambivalent moral forces in the human mind. The day 8 No'j is when to

Table 1.4. Phonetic symbols found in Colonial K'iche'an orthography.

Symbol	Documents	Modern	Phonetic
tresillo	ε	q	[q]
cuatrillo	4	k'	[k']
cuatrillo with *h*	4h	ch'	[č']
cuatrillo with cedilla	4,	tz'	[c']
k	k	q'	[q']

ask for good thoughts (Bunzel 1952:279; Edmonson 1997:121; Schultze Jena 1946:37; Tedlock 1992:122).

- *Tijax*, good day, the same as the preceding (Hernández Spina 1854). The day of quarrels and evil words. The day 8 Tijax is a good day to confess sins, especially quarrels with one's wife, relatives, or parents (Bunzel 1952:280; Edmonson 1997:121; Schultze Jena 1946:37; Tedlock 1992:122–123).

- *Kawoq*, indifferent day (Hernández Spina 1854); bad day, symbolic of evil embodied in the malice of the dead (Bunzel 1952:280; Edmonson 1997:121; Schultze Jena 1946:37; Tedlock 1992:123–124).

NOTES ON THE TRANSLATIONS AND TRANSCRIPTIONS

The following chapters present English translations from the original K'iche', Kaqchikel, or Spanish texts of calendar manuscripts. The texts are given in parallel columns with the transcription of the manuscript text in the left column and the English translation in the right. Word boundaries and line arrangement of the manuscript text have been adjusted and orthographic abbreviations are indicated in round brackets (—). Otherwise the transcription preserves the original orthography, which is based on Colonial Spanish spelling conventions. An additional five special characters were developed in the sixteenth century by Fr. Francisco de la Parra to designate glottalized consonants not found in Spanish. These symbols are given in Table 1.4.

Place-names are modernized to conform to current Instituto Geográfico Nacional standards (e.g., "Nebah" to "Nebaj"; "Ixtlavacan" to "Ixtahuacan"), and Spanish diacritics have been eliminated.

The major difficulty with any translation from an indigenous language is the reconstruction of original forms from unstandardized orthography. Although Karl Hermann Berendt and Vicente Hernández Spina were very faithful transcribers, they made some errors or copied misspellings from the original calendar manuscripts. Some of these transcription errors are straightforward and have been corrected in the text without further comment. All instances in which the orthography allows for more than one interpretation are discussed in the notes. Reconstructions, additions, and expansions of the text are indicated in brackets [—].

English language translations are given for all K'iche' or Kaqchikel text, with the exception of day and month names, as well as place-names. The meaning of day and month terms often involves multiple layers of meaning that would not be adequately reflected in an English translation. Similarly, the translation of place-names would hinder the recognition of the actual places. Likely translations of day and month names as well as place-names are therefore given as notes.

Some of the translations may appear to be flat prose. However, like most oral peoples the Colonial highland Maya depended on imagery to express their thoughts; therefore, the English translations may be filled with partially understood graphic similes and metaphors.

All K'iche'an terms within the English translation as well as analytical forms in the annotations are given in the standardized orthography for Mayan languages preferred by the Academia de las Lenguas Mayas de Guatemala. The use of arrowed brackets <—> in the annotations indicates that a referenced form preserves its original spelling.

Calendario de los indios de Guatemala, 1685

An important volume comprising transcriptions of two original documents is curated in the collections of the library at the University of Pennsylvania Museum of Archaeology and Anthropology. These manuscripts are identified as "Calendario de los indios de Guatemala 1685. Kaqchikel" and "Calendario de los indios de Guatemala 1722. Kiche." Both transcriptions were prepared in Guatemala City between 1875 and 1878 by the German philologist Carl Hermann Berendt. Daniel Garrison Brinton, who later acquired Berendt's manuscript collection, describes the two manuscripts: "Two precious pieces beautifully copied in facsimile by Dr. Berendt from ancient manuscripts he discovered in Guatemala. They present a detailed explanation of the calendars of the two nations, and may perhaps be the means of solving the strange problems presented by the chronology of the Mexican and Central American nations."

The first, a 1685 Kaqchikel calendar, was recorded by an anonymous Franciscan priest who wrote the "Crónica Franciscana," a manuscript that was at one time in the Franciscan convent in Guatemala City (Rodriguez and Crespo 1957:17). It was found in 1829 by Juan Gavarrete, a Guatemalan historian, among the books and papers of the convent, then deposited in the Archivo Arzobispal, and later returned to the Franciscans who showed it to the French abbé and scholar Charles Etienne Brasseur de Bourbourg. When it was returned to Gavarrete to be given to the Franciscans, their convent was no longer in existence and it was stored with manuscripts of the Sociedad Económica. In 1887, Berendt made a copy of folios 21–25 in the manuscript, which contained the calendar. This calendar was transcribed and published by Raquel Rodríguez and Mario Crespo in *Antropología e historia de Guatemala* (1957).

We do not know the origin of this calendar, although it probably derives from a Kaqchikel community near Santiago, Guatemala, most likely San Pedro La Laguna, a village located on the shoreline of Lake Atitlán in the department of Sololá: "In this town [San Pedro La Laguna] another calendar was found which is in my possession, stated the curate, but I don't understand it. The ones who make them are the same who are superstitious curers; it is almost impossible to identify them because of the great secrecy they keep about this; on certain occasions when I take and hide them there is such a furor that I release them again; the priest of Samayac . . . often he asks about them in confession but everybody denies they exist" (Cortés y Larraz 1958, 2:162).

It is clear from the text transcribed by Berendt that the Franciscan who copied it used a knowledgeable Kaqchikel as his informant (Carmack 1973:165). The 1685 calendar lists the eighteen months and the five additional days, indicating coefficient and day from the 260-day count for each day of the month. These calendar-round dates are correlated with the Gregorian year, starting on January 31, 1685 (1 Iq' 1 Tacaxepual), and ending on January 30, 1686 (1 Kamey 5 Tz'api Q'ij). It should be noted that throughout the manuscript, the coefficient of the day from the 260-day count is shifted back by one day. La Farge (1934:115) notes that "12 Akbal, May 12th, 1685, minus 101 days equals 2 Ik, January, 31st, 1685, but the Calendario Cakchiquel gives 1 Ik, as the 1st of Tacaxepual on that date."

/1/

Calendario de los indios de Guatemala 1685. Cakchiquel.

Copiado en la Ciudad de Guatemala. Marzo 1878

/2/

Advertencia

Este calendario se encuentra en la "Choronica de la S. Provincia del Santissimo Nombre de Jesus de Guattemala," conocido bajo el nombre de Crónica Franciscana, MS. que fué del convento de Franciscanos en esta ciudad. Hallado por Don Juan Gavarrete entre los libros y papeles de los conventos que en el año de 1829 se habían llevado al archivo arzobispal, fue de vuelto á los Franciscanos quienes lo prestaron al Abate Brasseur de Bourbourg. Cuando este lo mandó á Don Juan Gavarrete para devolverlo á los Franciscanos, ya no existía su convento y está guardado ahora entre los MSS. de la Sociedad Económica, para la cual está sacando una copia Don Juan.

Es un folio encuadernado en pergamino sin rótula ó portada con 283 fojas útiles, algunas muy dañadas, escrito en dos volumenes. Se ignora el nombre del autor, quien escribió su obra por los años de 1685.

[Crónica Franciscana] contiene dos libros, el primero de 48 capítulos (fojas 1–141) que alcanzan el año de 1541 y el otro de 42 capítulos (fojas 142–283) concluyendo con el año de 1600. El autor tuvo á la vista un número de MSS. de indios que aprovecha en los primeros capítulos del libro primero. En el cap. VII (fojas 21–25) trata "de los /3/ Kalendarios que usaban los Indios deste Reyno de Quauhutemala, que en seguida extracto.

Calendar of the Guatemalan Indians, 1685. Kaqchikel;
Copied in Guatemala City, March 1878

Preface

This calendar is found in the "Chronica de la S. Provincia del Santissimo Nombre de Jesus de Guatemala," also known under the name "Crónica Franciscana," in the Franciscan convent of this city. Discovered by Don Juan Gavarrete between the books and papers of the convents that were taken to the archbishop's archive in 1829, it went back to the Franciscans who lent it to Abate Brasseur de Bourbourg. At the time when he sent it to Don Juan Gavarrete in order to return it to the Franciscans, the convent did not exist anymore, and the manuscript is now kept between the manuscripts of the Sociedad Económica, for whom Don Juan is preparing a copy.

The folio is bound in parchment without label or title page with 283 leaves, some of them heavily damaged, in two volumes. The name of the author who has written this work in the year 1685 is unknown.

[The Crónica Franciscana] contains two books, the first of forty-eight chapters (folios 1–141), which extends to the year 1541, and the other one of forty-two chapters (folios 142–283), which ends with the year 1600. The author had a few Indian manuscripts at hand, which he made use of in the first chapters of the first book. In chapter 7 (folios 21–25) he deals with the calendars that the Indians of the Kingdom of Quauhutemala, as shows the following extract:

Dividian el año de la misma manera que los Mexicanos y solo se diferenciaban de ellos en los significados de los meses. Contaban el año de 18 meses y cada mes de veinte dias. A estos (360 dias) añadian cinco dias que decian validos (nemontemi de los Mexicanos) con que ajuntaron el año solar de 365 dias. Su primer dia o dia de año nuevo entre los Mexicanos era segun el P. Torquemada (Lib. X cap. 10 y 36) á primero de nuestro Febrero, aunque algunos dicen que solia comenzar á fines de Enero ó á principio de Marzo, y yo tengo por sin duda esto último porque como ni los Mexicanos ni estos alcanzaron el bisiesto— se apartaban y diferenciaban de nuestro calendario, y así ni estos, ni los Mexicanos comenzaban siempre su año a primero de nuestro febrero, sino que cada cuatro años se atrasaban un dia, v.g. el año de 1681, él de 82, él de 83 y él de 84 comenzó el año de los indios de este Reyno á primero de Febrero y este de 1685 comenzará á 31 de Enero y el de 1805 comenzará á primero de Enero y de ahí á 4 años á 31 de Diciembre etc.

Llamaban á su primero mes los Guatemaltecas Tacaxepual, que en su intelegencia de ellos quiere decir: tiempo de sembrar las primeras milpas: y si bien se advierte todo cuanto hacian y decian, era en orden al maiz, que poco faltó para tenerlo /4/ por Dios, y era, y es tanto el encanto y embelezo (?) que tienen con las milpas que por ellas olvidan hijos y mujer y otro cualquiera deleite, como se fuera la milpa su último fin y bienaventuranza. Los Mexicanos llamaban al primero mes de su año. Atlacahualco, que quiere decir "penuria de aguas" y era en estos el nombre de los meses en orden a los sacrificios que hacian á los dioses: los Guatemaltecas no, sino en orden a sus sustentamientos.

Meses de los Mexicanos

Atlacahualco: penuria de aguas

Tlacaxipeualiztli: fuellamiento de hombres

Tozoztontli: vigilia pequeña

Hueytozoztli: vigilia mayor

Toxcatl: deslizadero, resbalador

Etzalqualiztli: comida de poleadas

Tecuilhuitontli: pequeña fiesta de señores

They divided the year in the same way as the Mexicans and only distinguished themselves from them in the meaning of the months. They counted eighteen months of the year and every month of twenty days. To these (360 days) they added five days that they called "powerful" (the *nemontemi* of the Mexicans), which completed the 365 days of the solar year. The first day or new year's day among the Mexicans was according to Father Torquemada (Book 10, chapters 10 and 36) the first of our February, although some say that [the year] usually begins at the end of January or the beginning of March; and I am without doubt about this last one, because neither the Mexicans nor these [Guatemalans] invented the leap year. They differed and distinguished themselves from our calendar, and like this neither they or the Mexicans began their year on the first of our February, but every four years they deleted one day; for example, in the year of 1681, of 1682, of 1683, and of 1684 the year of the Indians of this kingdom began on the first of February and that [year] of 1685 began on January 31 and that of 1805 began on the first of January, and from there for four years on December 31, etc.

The Guatemalans called their first month "Tacaxepual," which according to their understanding means "season of sowing the first milpas." And, indeed, one notes that everything that they did and said was related to the maize, to that extent that there is little difference from having [maize] as a god, and it was [their god]. And the pleasure and delight they have with their milpas is that intense that they forget sons and wives and whichever pleasure about it; as if the milpa was the ultimate aim and well-being. The Mexicans called the first month of their year "Atlacahualco," which means "shortage of water." And among them, the names of the months were in the order of the offerings they made for the gods; whereas for the Guatemalans, they were in order of their [types] of sustenance.

Mexican Months

Atlacahualco: shortage of water

Tlacaxipeualiztli: flaying of men

Tozoztontli: small vigil

Hueytozoztli: main vigil

Toxcatl: tripping and slipping

Etzalqualiztli: eating of mush

Tecuilhuitontli: small ceremony of the lords

Hueitecuilhuitl: gran fiesta (de señores)

Tlaxuchimaco: repartimiento de flores

Xocotlhuetzi: se caen la frutas

Vchpaniztli: tiempo de escobas

Teotleco: llegada de los Dioses

Tepeilhuitl: fiesta de los montes

Quecholli: mes de aves pintados

Panquetzaliztli: enarbolmiento de pendones

Atemuztli: bajada de agua

Tititl: tiempo apretado

Izcalli: rescuitado

Meses de los Guatemaltecas

Tacaxepual: primera siembra

Nabeitumuzuz: tiempo de hormigas con alas

Rucantumuzuz: compañero del anterior

Cibixie: humo, quema de la broza

Vchum: tiempo de resembrar

Nabeimam: primer tiempo de revejecidos

Rucabmam: compañero del anterior

Lisinsá: tiempo de tierra blanda

Nabeitosis: primera cosecha de cacao

Rucatosis: segunda cosecha de cacao

Nabeipach: primer tiempo de empollar la clueca

Rucanpach: segundo tiempo de id.

Tziquinsih: tiempo de pájaros

Cakan: tiempo de celajas rojos

Ybota: tiempo de varios colores

Katic: pasante, o siembra comun

Yzcal: tiempo de retonos ó pimpollos

Pariche: tiempo de cobija, de frío.

Hueitecuilhuitl: great small ceremony of the lords

Tlaxuchimaco: distribution of flowers

Xocotlhuetzi: (when) the fruits fall

Vchpaniztli: season of brooms

Teotleco: arrival of the gods

Tepeilhuitl: ceremony of the mountains

Quecholli: month of colored birds

Panquetzaliztli: hoisting of flags

Atemuztli: downpour of water

Tititl: season of pressure

Izcalli: untying

Guatemalan Months

Takaxepual: first sowing

Nab'ey Tumuxux: season of flying ants

Rukab' Tumuxux: companion of the former

Qib'ixik: smoke, burning of brushwood

Uchum: season of second sowing

Nab'ey Mam: first season of the early-aged

Rukab' Mam: companion of the former

Likin Ka: season of soft earth

Nab'ey Toqik: first harvest of cacao

Rukab' Toqik: second harvest of cacao

Nab'ey Pach: first season of hen-hatching

Rukan Pach: second season of hen-hatching

Tz'ikin Q'ij: season of birds

Kaqan: season of red clouds

Ib'ota: season of various colors

Qatik: passing, general sowing

Iskal: season of sprouts and shoots

Pariche': season of blankets for the cold

/5/

Esto es lo que he podido saber y entender de sus Kalendarios: en cuanto á los sacrificios y ofrendas, no eran en estas gentes á la entrada de los meses como entre los Mexicanos, sino el dia que llamaban Vtzihohquih, buen dia, y fuese cuando fuese; de que los ha quedado tanta memoria, que hoy dia lo usan en muchas partes, y tienen buen día para tratos y contratos, buen dia para huirse, buen día para desafios.

Habia y hay entre los Indios algunos, que son maestros de esta arte diabólica, entre los Mexicanos se llamaban Tonalpouhqui y entre los Guatimaltecas Ahquih; el libro ó cuero en que tenian los caracteres ó signos de estas cuentas llamaban los Mexicanos tonalamatl y estos ɛamuh (apostella: chololqueh), y uno y otro quere decir "libro de suertes ó rollo de los días". Estos Ahquihes, como hombres que tienen pacto con el demonio, no solo encandilan á los demas indios haciendoles creer y tener por infalibles sus pronósticos y agueros, sino que como embaydores haciendolos perseverar en idolatrias, se hacen ellos adorar, trasformándose en sus nahuales en formas de animales feroces para hacerse temer.

Sucede con estas embaydores demas de anuncias á los pueblos los dias de Vtzilahquih, que guardan inviolables, y les salen como se los dicen; el que cuando les consueltan para saber alguna suerte ó cosa por venir, ó quien /6/ hizo mal ó un doliente; v.g. que el hacen ellos misteriosos argucandolas cejas, mirando al cielo suspirando y componiendo ciertas piedreguales, palitos y figuras, suele suceder tan á medida de lo que icen, que revelan lo que va escrito e una carta cerrada sin abrirla, lo que han hurtado, quien y como, lo que tiene en su corazon este ó el otro, estando ausentes, y otras mil cosas que menos que con diabólico auxilio sin imposibles de saberse, etc.

/8/		
Mes 1° Tacaxepual		
Tiempo de sembrar las primeras milpas.		
Dia 1	1 Yɛ	enero
		31

This is what I have managed to learn and understand about their calendars. With respect to the sacrifices and offerings, among these people, they did not take place at the beginning of the month as among the Mexicans, but on a day that they called Utzilaj Q'ij, "Good Day"; and that took place whenever [there happened to be such a day]. About this [concept of "good days"] they have preserved so much memory that today they use it in many parts, and they have good days for contracts, good days for fleeing, and good days for challenges.

There were and are some among the Indians who are masters of this diabolic art. Among the Mexicans they are called "Tonalpouhqui" and among the Guatemalans "Ajq'ij." The book or leather in which they have the symbols or signs of these counts is called "Tonalamatl" among the Mexicans, and among these Qamuj (pencil note: Cholol Q'ij), and one or the other may want to name them "book of fortunes" or "roll of the days." These Ajq'ijes, who are like men who have a pact with the devil, not only mislead the other Indians to believe and hold infallible the prognostication and omen but also make them preserve idolatry and make them worship [the idols] as deceivers by transforming into nahuals in the form of ferocious animals to frighten them.

With these deceivers, according to reports from the towns, in the days of Utzilaj Q'ij, which make them inviolable, whatever is said about them happens to them. One consults them to learn about fortune or something that might come, or who did evil, or a sufferer, for example. They make mysterious, suspicious gazes, observing the sky, breathing and arranging certain small stones, sticks, and figurines; it usually happens according to measure of what they say. They reveal what is written in a closed letter without opening it, what has been stolen, by whom and how, what somebody has in their heart, being absent, and a thousand other things that only with diabolic help would be possible to know, etc.

First Month Tacaxepoal[1]			
Season of sowing the first milpas.			
Day	1 Iq'		January
1			31

2	2 Aɛbal		febrero
			1
3	3 Kat		2
4	4 Can		3
5	5 Camey		4
6	6 Quieh		5
7	7 Kanel		6
8	8 Toh	buen dia	7
9	9 Tzij		8
10	10 Batz		9
11	11 Ee		10
12	12 Ah		11
13	13 Yiz		12
14	1 Tziquin	buen dia	13
15	2 Ahmak		14
16	3 Noh	buen dia	15
17	4 Tihax		16
18	5 Caok		17
19	6 Hunahpu	buen dia	18
20	7 Ymox		19

/9/

Mes 2° Nabeitumuzuz

Tiempo en que comienzan a volar hormigas on alas que es señal de primeras aguas.

Dia	8 Yɛ		febrero 20
1			
2	9 Aɛbal	vtzilahquih	21
3	10 Kat		22

2	2 Aq'ab'al		February
			1
3	3 K'at		2
4	4 Kan		3
5	5 Kamey		4
6	6 Kiej		5
7	7 Q'anel		6
8	8 Toj	good day	7
9	9 Tz'i'		8
10	10 B'atz'		9
11	11 E		10
12	12 Aj		11
13	13 I'x		12
14	1 Tz'ikin	good day	13
15	2 Ajmaq		14
16	3 No'j	good day	15
17	4 Tijax		16
18	5 Kawoq		17
19	6 Junajpu	good day	18
20	7 Imox		19

Second Month Nab'ey Tumuxux			
Season when the flying of winged ants begins, which is the sign of first water/ rains.			
Day	8 Iq'		20-Feb
1			
2	9 Aq'ab'al	good day	21
3	10 K'at		22

4	11 Can		23
5	12 Camey		24
6	13 Quieh		25
7	1 Kanel		26
8	2 Toh	vtzilahquih	27
9	3 Tzij		28
10	4 Batz	vtzilahquih	marzo 1
11	5 Ee	vtzilahquih	2
12	6 Ah	vtzilahquih	3
13	7 Yiz		4
14	8 Tziquin		5
15	9 Ahmak	vtzilahquih	6
16	10 Noh		7
17	11 Tihax		8
18	12 Caok		9
19	13 Hunahpu		10
20	1 Ymox		11

/10/

Mes 3° Rucantumuzuz

Hermano ó compañero del mes que buelan las hormigas.

Dia	2 Yε		marzo
1			12
2	3 Aεbal		13
3	4 Kat		14
4	5 Can		15
5	6 Camey		16
6	7 Quieh		17
7	8 Kanel		18

4	11 Kan		23
5	12 Kamey		24
6	13 Kiej		25
7	1 Q'anel		26
8	2 Toj	good day	27
9	3 Tz'i'		28
10	4 B'atz'	good day	1-Mar
11	5 E	good day	2
12	6 Aj	good day	3
13	7 I'x		4
14	8 Tz'ikin		5
15	9 Ajmaq	good day	6
16	10 No'j		7
17	11 Tijax		8
18	12 Kawoq		9
19	13 Junajpu		10
20	1 Imox		11

Third Month Rukan Tumuxux

Brother or companion of the month of the flying ants.

Day	2 Iq'	March
1		12
2	3 Aq'ab'al	13
3	4 K'at	14
4	5 Kan	15
5	6 Kamey	16
6	7 Kiej	17
7	8 Q'anel	18

8	9 Toh		19
9	10 Tzij		20
10	11 Batz		21
11	12 Ee		22
12	13 Ah		23
13	1 Yiz		24
14	2 Tziquin		25
15	3 Ahmak	vtzilahquih	26
16	4 Noh		27
17	5 Tihax	vtzilahquih	28
18	6 Caok		29
19	7 Hunahpu		30
20	8 Ymox		31

/11/

Mes 4° Cibixɛ

Tiempo de humo en que se siembra ó porque solian ser las quemas de la broza para sembrar ó por metaphora, nublado ó opaco por las nublazones que suele haber.

Dia	9 Yɛ		abril
1			1
2	10 Aɛbal		2
3	11 Kat		3
4	12 Can		4
5	13 Camey	vtzilahquih	5
6	1 Quieh		6
7	2 Kanel		7
8	3 Toh		8

8	9 Toj		19
9	10 Tz'i'		20
10	11 B'atz'		21
11	12 E		22
12	13 Aj		23
13	1 I'x		24
14	2 Tz'ikin		25
15	3 Ajmaq	good day	26
16	4 No'j		27
17	5 Tijax	good day	28
18	6 Kawoq		29
19	7 Junajpu		30
20	8 Imox		31

Fourth Month Qib'ixik

Season of smoke when one sows, or because there used to be the burning of brushwood before sowing, or as a metaphor, clouded or obscure because of the smog that there used to be.

Day	9 Iq'		April
1			1
2	10 Aq'ab'al		2
3	11 K'at		3
4	12 Kan		4
5	13 Kamey	good day	5
6	1 Kiej		6
7	2 Q'anel		7
8	3 Toj		8

9	4 Tzij		9
10	5 Batz	vtzilahquih	10
11	6 Ee		11
12	7 Ah		12
13	8 Yiz		13
14	9 Tziquin		14
15	10 Ahmak		15
16	11 Noh		16
17	12 Tihax	vtzilahquih	17
18	13 Caok		18
19	1 Hunahpu	vtzilahquih	19
20	2 Ymox		20

/12/

Mes 5° Vchum

Tiempo de resembrar.

Dia	3 Yɛ		abril
1			21
2	4 Aɛbal		22
3	5 Kat		23
4	6 Can		24
5	7 Camey		25
6	8 Quieh	vtzilahquih	26
7	9 Kanel	vtzilahquih	27
8	10 Toh		28
9	11 Tzij		29
10	12 Batz	vtzilahquih	30
11	13 Ee	vtzilahquih	mayo 1
12	1 Ah		2
13	2 Yiz		3
14	3 Tziquin		4

9	4 Tz'i'		9
10	5 B'atz'	good day	10
11	6 E		11
12	7 Aj		12
13	8 I'x		13
14	9 Tz'ikin		14
15	10 Ajmaq		15
16	11 No'j		16
17	12 Tijax	good day	17
18	13 Kawoq		18
19	1 Junajpu	good day	19
20	2 Imox		20

Fifth Month Uchum
Season of re-sowing.

Day	3 Iq'		April
1			21
2	4 Aq'ab'al		22
3	5 K'at		23
4	6 Kan		24
5	7 Kamey		25
6	8 Kiej	good day	26
7	9 Q'anel	good day	27
8	10 Toj		28
9	11 Tz'i'		29
10	12 B'atz'	good day	30
11	13 E	good day	1-May
12	1 Aj		2
13	2 I'x		3
14	3 Tz'ikin		4

15	4 Ahmak		5
16	5 Noh	vtzilahquih	6
17	6 Tihax	vtzilahquih	7
18	7 Caok		8
19	8 Hunahpu	vtzilahquih	9
20	9 Ymox		10

/13/

Mes 6° Nabeimam

Primer tiempo de revejidos, porque no crece muy alta la milpa que por este tiempo se sembraba, y aun las criaturas que nacian.

Dia	10 Yɛ		mayo
1			11
2	11 Aɛbal		12
3	12 Kat		13
4	13 Can	vtzilahquih	14
5	1 Camey		15
6	2 Quieh		16
7	3 Kanel		17
8	4 Toh		18
9	5 Tzij	vtzilahquih	19
10	6 Batz		20
11	7 Ee		21
12	8 Ah		22
13	9 Yiz		23
14	10 Tziquin		24
15	11 Ahmak		25
16	12 Noh	vtzilahquih	26

15	4 Ajmaq		5
16	5 No'j	good day	6
17	6 Tijax	good day	7
18	7 Kawoq		8
19	8 Junajpu	good day	9
20	9 Imox		10

Sixth Month Nab'ey Mam

First season of the early-aged, because the milpa sown in this season does not grow high, and likewise the creatures born in it.

Day	10 Iq'		May
1			11
2	11 Aq'ab'al		12
3	12 K'at		13
4	13 Kan	good day	14
5	1 Kamey		15
6	2 Kiej		16
7	3 Q'anel		17
8	4 Toj		18
9	5 Tz'i'	good day	19
10	6 B'atz'		20
11	7 E		21
12	8 Aj		22
13	9 I'x		23
14	10 Tz'ikin		24
15	11 Ajmaq		25
16	12 No'j	good day	26

17	13 Tihax		27
18	1 Caok		28
19	2 Hunahpu		29
20	3 Ymox	vtzilahquih	30

/14/

Mes 7º Rucabmam

Compañero del tiempo revejidos, por la misma razon.

Dia 1	4 Yε		mayo 31
2	5 Aεbal	vtzilahquih	junio 1
3	6 Kat		2
4	7 Can		3
5	8 Camey		4
6	9 Quieh	vtzilahquih	5
7	10 Kanel		6
8	11 Toh		7
9	12 Tzij		8
10	13 Batz		9
11	1 Ee		10
12	2 Ah		11
13	3 Yiz		12
14	4 Tziquin		13
15	5 Ahmak		14
16	6 Noh	vtzilahquih	15
17	7 Tihax		16
18	8 Caok		17
19	9 Hunahpu	vtzilahquih	18
20	10 Ymox		19

17	13 Tijax		27
18	1 Kawoq		28
19	2 Junajpu		29
20	3 Imox	good day	30

Seventh Month Rukab' Mam

Companion of the season of early-aged, for the same reason.

Day 1	4 Iq'		31-May
2	5 Aq'ab'al	good day	1-Jun
3	6 K'at		2
4	7 Kan		3
5	8 Kamey		4
6	9 Kiej	good day	5
7	10 Q'anel		6
8	11 Toj		7
9	12 Tz'i'		8
10	13 B'atz'		9
11	1 E		10
12	2 Aj		11
13	3 I'x		12
14	4 Tz'ikin		13
15	5 Ajmaq		14
16	6 No'j	good day	15
17	7 Tijax		16
18	8 Kawoq		17
19	9 Junajpu	good day	18
20	10 Imox		19

/15/

Mes 8° Liɛineá

Tiempo en que está la
tierra blanda y resbalosa
por las muchas aguas.

Dia	11 Yɛ		junio
1			20
2	12 Aɛbal		21
3	13 Kat		22
4	1 Can	vtzilahquih	23
5	2 Camey		24
6	3 Quieh	vtzilahquih	25
7	4 Kanel		26
8	5 Toh		27
9	6 Tzij		28
10	7 Batz		29
11	8 Ee	vtzilahquih	30
12	9 Ah	vtzilahquih	julio
			1
13	10 Yiz		2
14	11 Tziquin		3
15	12 Ahmak		4
16	13 Noh		5
17	1 Tihax		6
18	2 Caok		7
19	3 Hunahpu	vtzilahquih	8
20	4 Ymox		9

Eighth Month Likin Ka			
Season when the earth is soft and slippery because of the heavy rains.			
Day	11 Iq'		June
1			20
2	12 Aq'ab'al		21
3	13 K'at		22
4	1 Kan	good day	23
5	2 Kamey		24
6	3 Kiej	good day	25
7	4 Q'anel		26
8	5 Toj		27
9	6 Tz'i'		28
10	7 B'atz'		29
11	8 E	good day	30
12	9 Aj	good day	July
			1
13	10 I'x		2
14	11 Tz'ikin		3
15	12 Ajmaq		4
16	13 No'j		5
17	1 Tijax		6
18	2 Kawoq		7
19	3 Junajpu	good day	8
20	4 Imox		9

/16/

Mes 9° Nabeitoɛiɛ

Cosecha primera de
Cacao ó tiempo de
desollar

Dia	5 Yɛ	vtzilahquih	julio
1			10
2	6 Aɛbal	vtzilahquih	11
3	7 Kat		12
4	8 Can		13
5	9 Camey	vtzilahquih	14
6	10 Quieh		15
7	11 Kanel		16
8	12 Toh		17
9	13 Tzij		18
10	1 Batz		19
11	2 Ee		20
12	3 Ah		21
13	4 Yiz		22
14	5 Tziquin		23
15	6 Ahmak		24
16	7 Noh		25
17	8 Tihax	vtzilahquih	26
18	9 Caok		27
19	10 Hunahpu		28
20	11 Ymox		29

Ninth Month Nab'ey Toqik

First harvest of cacao or season of flaying.

Day	5 Iq'	good day	July
1			10
2	6 Aq'ab'al	good day	11
3	7 K'at		12
4	8 Kan		13
5	9 Kamey	good day	14
6	10 Kiej		15
7	11 Q'anel		16
8	12 Toj		17
9	13 Tz'i'		18
10	1 B'atz'		19
11	2 E		20
12	3 Aj		21
13	4 I'x		22
14	5 Tz'ikin		23
15	6 Ajmaq		24
16	7 No'j		25
17	8 Tijax	good day	26
18	9 Kawoq		27
19	10 Junajpu		28
20	11 Imox		29

/17/

Mes 10° Rucactoɛiɛ

Cosecha segunda
o compañero del
antecedente.

Dia	12 Yɛ		julio
1			30
2	13 Aɛbal		31
3	1 Kat		agosto
			1
4	2 Can		2
5	3 Camey		3
6	4 Quieh		4
7	5 Kanel	vtzilahquih	5
8	6 Toh		6
9	7 Tzij		7
10	8 Batz		8
11	9 Ee	vtzilahquih	9
12	10 Ah		10
13	11 Yiz		11
14	12 Tziquin		12
15	13 Ahmak		13
16	1 Noh		14
17	2 Tihax		15
18	3 Caok		16
19	4 Hunahpu		17
20	5 Ymox		18

Tenth Month Rukab' Toqik

Second harvest or companion of the former.

Day	12 Iq'		July
1			30
2	13 Aq'ab'al		31
3	1 K'at		August
			1
4	2 Kan		2
5	3 Kamey		3
6	4 Kiej		4
7	5 Q'anel	good day	5
8	6 Toj		6
9	7 Tz'i'		7
10	8 B'atz'		8
11	9 E	good day	9
12	10 Aj		10
13	11 I'x		11
14	12 Tz'ikin		12
15	13 Ajmaq		13
16	1 No'j		14
17	2 Tijax		15
18	3 Kawoq		16
19	4 Junajpu		17
20	5 Imox		18

/18/

Mes 11° Nabeypach

Primer tiempo de
empollar las cluecas.

Dia	6 Yɛ	vtzilahquih	agosto
1			19
2	7 Aɛbal		20
3	8 Kat		21
4	9 Can		22
5	10 Camey		23
6	11 Quieh	vtzilahquih	24
7	12 Kanel		25
8	13 Toh		26
9	1 Tzij		27
10	2 Batz		28
11	3 Ee	vtzilahquih	29
12	4 Ah		30
13	5 Yiz		31
14	6 Tziquin		septiembre 1
15	7 Ahmak	vtzilahquih	2
16	8 Noh	vtzilahquih	3
17	9 Tihax		4
18	10 Caok		5
19	11 Hunahpu		6
20	12 Ymox		7

Eleventh Month Nab'ey Pach			
First season of hen-hatching.			
Day	6 Iq'	good day	August
1			19
2	7 Aq'ab'al		20
3	8 K'at		21
4	9 Kan		22
5	10 Kamey		23
6	11 Kiej	good day	24
7	12 Q'anel		25
8	13 Toj		26
9	1 Tz'i'		27
10	2 B'atz'		28
11	3 E	good day	29
12	4 Aj		30
13	5 I'x		31
14	6 Tz'ikin		September 1
15	7 Ajmaq	good day	2
16	8 No'j	good day	3
17	9 Tijax		4
18	10 Kawoq		5
19	11 Junajpu		6
20	12 Imox		7

/19/

Mes 12° Rucanpach

Segundo tiempo de
empollar las cluecas.

Dia	13 Yε		septiembre
1			8
2	1 Aεbal		9
3	2 Kat		10
4	3 Can		11
5	4 Camey		12
6	5 Quieh		13
7	6 Kanel		14
8	7 Toh		15
9	8 Tzij		16
10	9 Batz	vtzilahquih	17
11	10 Ee		18
12	11 Ah		19
13	12 Yiz	vtzilahquih	20
14	13 Tziquin		21
15	1 Ahmak		22
16	2 Noh		23
17	3 Tihax	vtzilahquih	24
18	4 Caok		25
19	5 Hunahpu	vtzilahquih	26
20	6 Ymox	vtzilahquih	27

Twelfth Month Rukan Pach			
Second season of hen-hatching.			
Day	13 Iq'		September
1			8
2	1 Aq'ab'al		9
3	2 K'at		10
4	3 Kan		11
5	4 Kamey		12
6	5 Kiej		13
7	6 Q'anel		14
8	7 Toj		15
9	8 Tz'i'		16
10	9 B'atz'	good day	17
11	10 E		18
12	11 Aj		19
13	12 I'x	good day	20
14	13 Tz'ikin		21
15	1 Ajmaq		22
16	2 No'j		23
17	3 Tijax	good day	24
18	4 Kawoq		25
19	5 Junajpu	good day	26
20	6 Imox	good day	27

/20/

Mes 13° Tziquinɛih

Tiempo de pájaros.

Dia	7 Yɛ		septiembre
1			28
2	8 Aɛbal	vtzilahquih	29
3	9 Kat		30
4	10 Can	vtzilahquih	octubre
			1
5	11 Camey		2
6	12 Quieh		3
7	13 Kanel		4
8	1 Toh		5
9	2 Tzij		6
10	3 Batz	vtzilahquih	7
11	4 Ee	vtzilahquih	8
12	5 Ah		9
13	6 Yiz		10
14	7 Tziquin	vtzilahquih	11
15	8 Ahmak		12
16	9 Noh	vtzilahquih	13
17	10 Tihax		14
18	11 Caok		15
19	12 Hunahpu		16
20	13 Ymox		17

Thirteenth Month Tz'ikin Q'ij			
Season of birds.			
Day	7 Iq'		September
1			28
2	8 Aq'ab'al	good day	29
3	9 K'at		30
4	10 Kan	good day	October
			1
5	11 Kamey		2
6	12 Kiej		3
7	13 Q'anel		4
8	1 Toj		5
9	2 Tz'i'		6
10	3 B'atz'	good day	7
11	4 E	good day	8
12	5 Aj		9
13	6 I'x		10
14	7 Tz'ikin	good day	11
15	8 Ajmaq		12
16	9 No'j	good day	13
17	10 Tijax		14
18	11 Kawoq		15
19	12 Junajpu		16
20	13 Imox		17

/21/

Mes 14° Cakan

Tiempo de colores rojas
y de flores amarillas

Dia	1 Yɛ	octubre
1		18
2	2 Aɛbal	19
3	3 Kat	20
4	4 Can	21
5	5 Camey	22
6	6 Quieh	23
7	7 Kanel	24
8	8 Toh	25
9	9 Tzij	26
10	10 Batz	27
11	11 Ee	28
12	12 Ah	29
13	13 Yiz	30
14	1 Tziquin	31
15	2 Ahmak	noviembre 1
16	3 Noh	2
17	4 Tihax	3
18	5 Caok	4
19	6 Hunahpu	5
20	7 Ymox	6

Fourteenth Month Kaqan		
Season of red colors and yellow flowers		
Day	1 Iq'	October
1		18
2	2 Aq'ab'al	19
3	3 K'at	20
4	4 Kan	21
5	5 Kamey	22
6	6 Kiej	23
7	7 Q'anel	24
8	8 Toj	25
9	9 Tz'i'	26
10	10 B'atz'	27
11	11 E	28
12	12 Aj	29
13	13 I'x	30
14	1 Tz'ikin	31
15	2 Ajmaq	November 1
16	3 No'j	2
17	4 Tijax	3
18	5 Kawoq	4
19	6 Junajpu	5
20	7 Imox	6

/22/

Mes 15° Ybotá

Tiempo de varias colores rojas ó de arrollar petates.

Dia	8 Yε	noviembre
1		7
2	9 Aεbal	8
3	10 Kat	9
4	11 Can	10
5	12 Camey	11
6	13 Quieh	12
7	1 Kanel	13
8	2 Toh	14
9	3 Tzij	15
10	4 Batz	16
11	5 Ee	17
12	6 Ah	18
13	7 Yiz	19
14	8 Tziquin	20
15	9 Ahmak	21
16	10 Noh	22
17	11 Tihax	23
18	12 Caok	24
19	13 Hunahpu	25
20	1 Ymox	26

Fifteenth Month Ib'ota'

Season of various red colors or of rolling up mats.

Day	8 Iq'	November
1		7
2	9 Aq'ab'al	8
3	10 K'at	9
4	11 Kan	10
5	12 Kamey	11
6	13 Kiej	12
7	1 Q'anel	13
8	2 Toj	14
9	3 Tz'i'	15
10	4 B'atz'	16
11	5 E	17
12	6 Aj	18
13	7 I'x	19
14	8 Tz'ikin	20
15	9 Ajmaq	21
16	10 No'j	22
17	11 Tijax	23
18	12 Kawoq	24
19	13 Junajpu	25
20	1 Imox	26

/23/

Mes 16° Katic		
Pasante ó siembra común.		
Dia	2 Yɛ	noviembre
1		27
2	3 Aɛbal	28
3	4 Kat	29
4	5 Can	30
5	6 Camey	diciembre 1
6	7 Quieh	2
7	8 Kanel	3
8	9 Toh	4
9	10 Tzij	5
10	11 Batz	6
11	12 Ee	7
12	13 Ah	8
13	1 Yiz	9
14	2 Tziquin	10
15	3 Ahmak	11
16	4 Noh	12
17	5 Tihax	13
18	6 Caok	14
19	7 Hunahpu	15
20	8 Ymox	16

Sixteenth Month Qatik
Passing or general sowing.

Day	2 Iq'	November
1		27
2	3 Aq'ab'al	28
3	4 K'at	29
4	5 Kan	30
5	6 Kamey	December 1
6	7 Kiej	2
7	8 Q'anel	3
8	9 Toj	4
9	10 Tz'i'	5
10	11 B'atz'	6
11	12 E	7
12	13 Aj	8
13	1 I'x	9
14	2 Tz'ikin	10
15	3 Ajmaq	11
16	4 No'j	12
17	5 Tijax	13
18	6 Kawoq	14
19	7 Junajpu	15
20	8 Imox	16

/24/

Mes 17° Yzcal

Que es retoñar o echar pimpollas.

Dia	9 Yɛ	diciembre
1		17
2	10 Aɛbal	18
3	11 Kat	19
4	12 Can	20
5	13 Camey	21
6	1 Quieh	22
7	2 Kanel	23
8	3 Toh	24
9	4 Tzij	25
10	5 Batz	26
11	6 Ee	27
12	7 Ah	28
13	8 Yiz	29
14	9 Tziquin	30
15	10 Ahmak	31
16	11 Noh	1 [enero] 1686
17	12 Tihax	2
18	13 Caok	3
19	1 Hunahpu	4
20	2 Ymox	5

Seventeenth Month Iskal[2]

That which is sprouting or to throw buds.

Day	9 Iq'	December
1		17
2	10 Aq'ab'al	18
3	11 K'at	19
4	12 Kan	20
5	13 Kamey	21
6	1 Kiej	22
7	2 Q'anel	23
8	3 Toj	24
9	4 Tz'i'	25
10	5 B'atz'	26
11	6 E	27
12	7 Aj	28
13	8 I'x	29
14	9 Tz'ikin	30
15	10 Ajmaq	31
16	11 No'j	[January] 1, 1686
17	12 Tijax	2
18	13 Kawoq	3
19	1 Junajpu	4
20	2 Imox	5

/25/

Mes 18° Pariche

Tiempo de cobija para guardarse del frio.

Dia	3 Yε	enero
1		6
2	4 Aεbal	7
3	5 Kat	8
4	6 Can	9
5	7 Camey	10
6	8 Quieh	11
7	9 Kanel	12
8	10 Toh	13
9	11 Tzij	14
10	12 Batz	15
11	13 Ee	16
12	1 Ah	17
13	2 Yiz	18
14	3 Tziquin	19
15	4 Ahmak	20
16	5 Noh	21
17	6 Tihax	22
18	7 Caok	23
19	8 Hunahpu	24
20	9 Ymox	25

Eighteenth Month
Pariche'

Season of blankets to protect oneself against the cold.

Day	3 Iq'	January
1		6
2	4 Aq'ab'al	7
3	5 K'at	8
4	6 Kan	9
5	7 Kamey	10
6	8 Kiej	11
7	9 Q'anel	12
8	10 Toj	13
9	11 Tz'i'	14
10	12 B'atz'	15
11	13 E	16
12	1 Aj	17
13	2 I'x	18
14	3 Tz'ikin	19
15	4 Ajmaq	20
16	5 No'j	21
17	6 Tijax	22
18	7 Kawoq	23
19	8 Junajpu	24
20	9 Imox	25

/26/			
Tzapi εih			
Puerta que cierra el año, dia y tiempo.			
Dia 1	10 Yε	enero 26	
2	11 Aεbal	27	
3	12 Kat	28	
4	13 Can	29	
5	1 Camey	30	

Closing Days			
Door that closes the year, day, and season.			
Day 1	10 Iq'	26-Jan	
2	11 Aq'ab'al	27	
3	12 K'at	28	
4	13 Kan	29	
5	1 Kamey	30	

THREE

Calendario de los indios de Guatemala, 1722

The second manuscript, "Calendario de los indios de Guatemala 1722. Kiche," was written in K'iche' in 1722 and was copied by Karl Hermann Berendt in 1877 from a manuscript in the Museo Nacional de Guatemala. Berendt believed that the calendar came from the Quetzaltenango area and that it was the one given to Cortés y Larraz by the resident priest (Carmack 1973:166).

The 1722 K'iche' calendar is the most complex and interesting of the three manuscripts in this volume. It consists of nine parts arranged in three sections indicated by A, B, and C. Individual annual cycles within these three calendars are indicated by Roman numerals (I–IX), followed by pagination in the original manuscript. The brief summary in Table 3.1 indicates the calendar type (i.e., solar, 260-day count, etc.), the beginning and ending points of each calendar, and, if available, the correlation with the Gregorian date.

Edmonson (1997) produced a nearly complete translation and analysis of the calendars (*chol q'ij*) and almanacs (*ajilab'al q'ij*) from the Berendt manuscript, em-

Table 3.1. Composition of Calendario de los indios de Guatemala, 1722.

	ff.	
	i–xi	An introduction by Carl Hermann Berendt including excerpts from the *Historia de la provincia de San Vicente de Chiapa y Guatemala* by Francisco Ximénez and the *Descripción geográfico moral de la dioceses de Goathemala* by Pedro Cortés y Larraz regarding the Guatemalan indigenous calendrical system as well as a drawing of a calendar wheel of the 20-day ritual calendar that may have been added by Berendt.
A		Chol poal q'ij, macehual q'ij.
A-I	1–18	Correlation of a complete 365-day extract (solar year) of the 18,980-day calendar indicating dates and weekdays (dominical letters) in the Gregorian calendar. Begins on 9 Keej 20 Nab'e Mam and ends on 11 Aj (May 3, 1722–May 4, 1723).
A-II	18–19	Correlation of an abridged 365-day extract (solar year) of the 18,980-day calendar indicating dates in the Gregorian calendar. Begins on 10 E 20 Nab'e Mam and ends on 11 No'j (May 4, 1723–May 4, 1724).
A-III	19–20	Abridged 365-day extract (solar year) of the 18,980-day calendar indicating dates in the Gregorian calendar. Begins on 11 No'j 20 Nab'e Mam and ends on 12 Iq' (1724–1725).
A-IV	20–21	Abridged 365-day extract (solar year) of the 18,980-day calendar indicating dates in the Gregorian calendar. Begins on 12 Iq' 20 Nab'e Mam and ends on 13 Keej (1725–1726).
A-V	21	Abridged 365-day extract (solar year) of the 18,980-day calendar indicating dates in the Gregorian calendar. Begins on 13 Keej, 20 Nab'e Mam and ends on 1 E (1726–1727).
A-VI	22	Abridged 365-day extract (solar year) of the 18,980-day calendar indicating dates in the Gregorian calendar. Begins on 1 E, 20 Nab'e Mam and ends on 2 No'j (1726–1727).
B-I	23–36	Ajilab'al q'ij
B-II	37–49	A complete 260-day cycle (see B-I). The count starts on 1 Q'anil (day 248) and ends on 13 Keej (day 247).
C-I	50	A list of the eighteen month names including the five closing days, Tz'api Q'ij; begins with the month Che' and ends with Rox Si'j.

phasizing the significance of the almanacs and their relation to modern divinatory practice. His earlier study of Mesoamerican calendars (1988) includes comments on the general calendrics of the 1722 K'iche' calendar.

CALENDAR A

On pages 1 through 23 is an extract of six consecutive years, 1722 to 1727, from a calendar round. Each of the six calendars correlates the 260-day count and the

Table 3.2. Correction of error in Calendar A-I.

Days	Gregorian	Calendar A-I	Correction	Calendar A-II
361	April 28, 1723	5 Keej 20 K'isb'al Rech		
362	April 29, 1723	6 Q'anil 1 K'isb'al Rech		
363	April 30, 1723	7 Toj 2 K'isb'al Rech		
364	May 1, 1723	8 Tz'i'		
365	May 2, 1723	3 K'isb'al Rech 9 B'atz'		
1	May 3, 1723	4 K'isb'al Rech 10 E	10 E	
2	May 4, 1723	5 K'isb'al Rech 11 Aj	20 Nab'e Mam	10 E 20 Nab'e Mam

eighteen K'iche' month names, as well as the five closing dates of the 365-day solar year, including information about the four year bearers' names, with the Gregorian calendar for the years 1722 to 1727 (Carmack 1973:166; Edmonson 1997:115).

In Calendar A-I, the last five days of the preceding year mentioned are 4 Iq', 5 Aq'ab'al, 6 K'at, 7 Kan, 8 Kame, followed by the four year bearers. On pages 1–18, the author lists all of the days of the full solar year that begins on the date 9 Keej 20 Nab'e Mam, corresponding to May 3, 1722. Furthermore, the first day of each of the eighteen solar months as well as the first day of the five closing days are given, thus indicating the calendar-round dates. For each calendar-round position between May 3, 1722, and May 4, 1723, the calendar notes the exact Gregorian date. There are further marks of dates with commentaries about agriculturally relevant activities, such as the beginning of the planting or harvesting seasons.

Irregularities at the end of Calendar A-I affect the length of the K'iche' solar year (1722–23). The author of Calendar A made a mistake when indicating the final five days of the year, the Tz'api Q'ij or K'isb'al Rech (Table 3.2). In Calendar A-I the Wayeb days begin with the calendar-round position 5 Keej, 20 K'isb'al Rech, which correlates to April 28, 1723. The solar year should end five days later on 4 B'atz, 4 K'isb'al Rech (May 2, 1723). As a consequence, the new K'iche' solar year should begin on 10 E 20 Nab'e Mam (May 3, 1723). Indeed, Calendar A-II begins correctly on 10 E 20 Nab'e Mam, which however is correlated with May 4, 1723. The author mistakenly ended the preceding year (A-I) on the calendar-round position 10 E 5 K'isb'al Rech and thus added a sixth day to the Wayeb count (May 3, 1723).

The addition of an extra day shifts the beginning of the new year in the Gregorian calendar a day forward. The explanation for this error may be found in the notation of the first day, or "completion day," of the month. This is the first day of the month although, arithmetically, it is day zero, "the seating" of the new month, or the last, or twentieth, day of the past month (Thompson 1950:121). The author evidently knew the concept of Wayeb but misinterpreted it and added five days to the first position of K'isb'al Rech, thus extending this period to six instead of five.

Calendar A-I correlates each single calendar-round position of the indigenous solar year with the corresponding Gregorian date and with the corresponding weekday by means of dominical letters. In the European tradition of the perpetual calendar (*calendarium perpetuum*), the days of the year are marked with the letters *A, b, c, d, e, f,* and *g* in a cycle of 7: January 1 of each year is marked with the capital letter *A*, January 2 with the letter *b*, . . . January 7 with the letter *g*, January 8 with the letter *A* again, and so forth. The letters were used to determine the Sundays, and especially Easter Sunday, of the year. The letter of the first Sunday is the dominical letter. If, for instance, in a given year, the first Sunday falls on January 4, the dominical letter of the year would be d. Accordingly, the days of the week would be: Monday (*e*), Tuesday (*f*), Wednesday (*g*), Thursday (*A*), Friday (*b*), Saturday (*c*), and Sunday (*d*).

Regarding the numeration by dominical letters, we have to point out another error in Calendar A-I. December 31, 1722, is indicated by the letter *A*. However, the author continues to count with the letter *b* on January 1, 1723, instead of starting over with *A*. The incorrect correlation of dominical letters continues through the entire calendar. The dominical letter of the year 1722 is *d* according to the Gregorian count.

In Calendar A-I all occurrences of the day Keej and the numeral 13 are marked with an asterisk (*). Similarly, 13 Keej is marked with two asterisks (**). The day and the month name of each calendar round in Calendar A-I are underlined with red ink. It is likely that these marks (i.e., asterisks and underlinings) were not part of the original manuscript transcribed by Berendt in Guatemala but were added by him.

Continuing the count of Calendar A-I, Calendars A-II to A-VI are abbreviated in that they only give calendar-round positions for the first day of each of eighteen months and the five Wayeb days. Calendars A-II and A-IV indicate the Gregorian dates for each of the months, but Calendars A-III, A-V, and A-VI omit this information. Calendar A-IV (1725–26) indicates the beginning of the month Che', that is, the calendar round 12 Iq', 20 Che', as falling on January 13.

Several annotations and corrections to the Gregorian dates in Calendar A-II have been added by a different hand, presumably by Daniel Garrison Brinton, who purchased Berendt's manuscript collection. Brinton seems to have observed that Calendar A-II only indicates twelve days of difference for the calendar-round position 13 E, 20 Tz'isi Laqam (September 1, 1723) and 7 E, 20 Tz'ikin Q'ij (September

Table 3.3. Corrections for Gregorian equivalents in Calendar A-II.

Berendt transcription	Gregorian date	Transcript corrections	Corrections based on Edmonson
10 E 20 Nabema	May 4		May 3
4 E 20 Vcab Ma	May 23		
11 E 20 Liquinca	June 12		
5 E 20 Vcab Liquinca	July 2		
12 E 20 Nabe Pach	July 22		
6 E 20 Ucab Pach	August 12	August 11	August 11
13 E 20 4,içilakam	September 1	August 31	August 31
7 E 20 4,iquin εih	September 12	September 20	September 21
1 E 20 Cakam	October 12	October 10	October 11
8 E 20 Botam	November 1	October 30	October 31
2 E 20 Nabeçih	November 19		November 20
9 E 20 Vacabçih	December 9		December 10
3 E 20 Roxçih	December 29		December 30
10 E 20 Chee	January 18		January 19
4 E 20 Tequexepual	February 7		February 8
11 E 20 4,ibap[o]p	February 27		February 28
5 E 20 Çac	March 20	March 19	March 19
12 E 20 4hab	April 9		April 8
6 E 20 4izo			April 28

12, 1723), when the arithmetic difference between the single Maya months should be twenty days. The annotation thus adds the correct date, that is, September 20. Brinton also corrected the error in the calculation of the length of the year repeated in Calendar A-II. Table 3.3 lists the correlations of calendar-round and Gregorian dates and also the corrections found in Berendt's transcription or suggested by Edmonson (1997:119).

The solar year in Calendar A begins with the first day of Nab'e Mam. The tzolk'in notation of this day is the year bearer, referred to as "Mam" in modern K'iche' (Tedlock 1992:89). There are four year bearers in the Maya calendar. The least common multiple of the permutation of the 260-day and the 365-day cycle is 5, the same combination of tzolk'in day name and coefficient of the haab-month, and thus only reoccurs every four years. This means that there are five groups of four days. The group of days falling on the Maya new year's days are the year bearers. The K'iche' calendar maintains the Classic year bearers: Ik', Manik, Eb, and Caban (Iq', Keej, E, and No'j; Tedlock 1992:91–92). There still exist rituals associated with the year bearers, and the numeral that precedes the day sign still indicates the quality attributed to the solar year (Bunzel 1952:283).

The manuscript does not offer information on the beginning of the chol q'ij, or ritual calendar. Tedlock (1992:96) has proposed that each K'iche' community had its own beginning of the 260-day count. In Appendix 1, Burkitt, however, argues that there is much variability in the placement of the beginning of the calendar.

CALENDAR B

The second part of the manuscript consists of two 260-day almanacs or divinatory calendars (*ajilab'al q'ij*), the first dating to 1722 (pages 23–36) and the second one probably dating to 1770 (pages 37–49). Berendt indicates that both almanacs are written by more than one hand. There are indications that the two almanacs derive from different places, as they seem to exhibit significant dialectal variability (e.g., *qa tzij* <katzih>—*qi tzij* <kitzih>, "truly"; *ka'ib'* <caib, cayb>—*keb'* <queb, queeb>, "two, twice"; *q'ojon* <ɛohon>—*q'ojom* <ɛohom>, "drum") and deviate in the use of individual day names (e.g., "Yix" and "Balam"), orthographic realization (e.g., "Imox" and "Imos"), as well as in their specific forecasts for individual days (Schuller n.d.:4, 29). Also, the style of the individual prognostications differs in both calendars; for example, the first author ends every prognostication with the phrase *ro'ichal* <ro yichal> (all five of them), indicating that this forecast applies to all five days. The second divinatory calendar exhibits less standardized orthography than the first one. The dialectal variability suggests that the second divinatory calendar was produced in the Quetzaltenango area, whereas the first probably originates from the central region.

The two almanacs are special in that the days are organized in groups of five that share the same coefficient and forecast (Carmack 1973:166). The vertical distance between two sequential days within a single group is fifty-two days, whereas the distance between the groups is a single day. As noted elsewhere (La Farge 1947:180–181; Edmonson 1997), the unusual layout is reminiscent of the Codex Dresden; however, this cannot be taken as an indication for any hieroglyphic antecedent of the 1722 divinatory calendar.

The first almanac begins on 1 Imox and ends on 13 Junajpu, the second begins on 1 Q'anil and ends on 13 Keej. A Gregorian calendar date is noted on page 28 (5 Q'anil, or March 13, 1770). Otherwise the almanacs are not meant to be correlated with the Christian year.

The significance of a small drawing of a crosshatched circle on page 49 adjacent to a group of five days that begin with 11 Imox is unknown. It is also uncertain whether Berendt transcribed the drawing from the original manuscript or added it from his own familiarity with Mayan hieroglyphic writing.

CALENDAR C

The final calendar on page 50 lists the nineteen month names of the K'iche' calendar and notes erroneously that the year had 400 days. The end of the year is indicated after the month Urox Si'j. The Tz'api Q'ij have been inserted between the month names Ukab' Mam and Likinka. Thus, Calendar C indicates a second end of the year. The scribe evidently did not properly understand the arithmetic involved in the calendar. Calendar C lists the months of the year in a different order than Calendar A, in which the final five days are inserted between the months Ch'ab' and Nab'e Mam. This suggests that the individual calendars within the 1722 manuscript derive from other sources.

CALENDAR CORRELATION

The 1722 K'iche' calendar is strong evidence for the use of written calendar-round dates in Guatemala. In Calendars A and B, calendar-round dates are correlated with Gregorian dates. This document therefore may serve to test suggested correlations for the Classic Maya calendar with the Christian calendar. A review of the relevant literature indicates that the three K'iche' calendars copied by Berendt have been peripheral to the correlation question thus far.

The correlation of the ancient Mesoamerican calendar with the Gregorian calendar has been an issue of some debate among Mayanists for a long time (Teeple 1931; Thompson 1950, 1971) but is considered to be resolved today (Lounsbury 1978). The 11.16.0.0.0-correlation of Joseph Goodman, J. Eric S. Thompson, and Juan Martínez Hernández is the generally accepted correlation, as it accommodates all historical, epigraphic, and astronomical criteria and has been confirmed by radiocarbon dating (Brack-Bernsen 1977:27; Lounsbury 1978, 1992; Satterthwaite and Ralph 1960).

The 1722 K'iche' calendar provides additional support for Thompson's revised correlation constant of 584,283 days. Testing various suggested constants for the correlation of the Classic Maya and the Christian calendars on the basis of the correlation dates given in the 1722 K'iche' calendar, the GMT-584,283-correlation is the only correlation that meets several calculatory criteria.

The fact that the 1722 K'iche' calendar provides support for the otherwise well attested and confirmed GMT-584,283-correlation suggests that the 260-day and the 365-day count of the prehispanic calendar continued without interruption or break well into Colonial times. Thompson already made this same argument based on a correlation date from Tovilla's 1635 *Relación histórica descriptiva de la provincias de la Verapaz y de la del Manché* (1960) and contemporary calendar usage among the Ixil, concluding that although the sequence of the month names in highland Guatemala shifted, the calendar itself continued without break since Classic Maya times (Thompson 1950:310).

Thompson (1934, 1950) did not include the evidence from the 1722 K'iche' calendar in the arguments for his correlation. Although he makes reference to the 1722 K'iche' manuscript (Thompson 1950:68), he apparently did only refer to its citation in a contribution by Oliver La Farge (1934). Munro Edmonson (1988), in his study of Mesoamerican chronology, assumes the GMT-584,283-correlation for the 1722 calendar as given but does not provide calculatory proof. For a detailed description and mathematical proof of the correlation constant in the K'iche' manuscript, we refer the reader to Appendix 2.

/i/ Calendario de los indios de Guatemala 1722 Kiché. Copiado en la Ciudad de Guatemala, Abril 1877.

/ii/

Advertencia

El original de este calendario en lengua Kiché, erroneamente llamado Calendario Kachiquel en el Catálogo de la biblioteca de la sección etnológica del Museo Nacional (Guatemala, 1875) pag. 8 no. 1, forma un cuaderno de 24 fojas útiles en 4to menor. Se halla el prinicipio de un volumen en folio, intitulado "Larras, Opúsculos."

La segunda parte, comenzando en la página 37 de la presente copia está escrita por otra mano con ortografía diferente, trayendo al márgen (página 38 de esta copia) la fecha 13 de Mayo de 1770 años.

Parece que es el mismo calendario del cual habla el Arzobispo D. Pedro Cortés y Larraz en su "Descripción Geográfico-Moral de la Diócesis de Guathemala hecho en la visita que hizo de ella en los /iii/ años de 1768, 1769 y 1770, foja 142 vuelta, diciendo:

> En el Pueblo de Quesaltenango encontré un Kalendario de ellos, que deseaba mucho llegara a mis manos y otro hallé en el de San Pedro de la Laguna, escrito ue un Indio en el año 1732 de muy buena letra en el Idioma Kiché; los he manifestado á los mejores Maestros de dicho Idioma, y aunque saben el significado de los términos, pero ninguno penetra su concepto, á excepcion de su último apartado, en que está puesto con claridad bastante, que aquel dia era oportuno para hacer cantar responsos, celebrar misas, y otras funciones devotas. Entiendo que entre sus papeles [de los Indios] se encontrarían en estos pueblos raras historias del Rey de Kiché, porque estos Indios tienen (á mi parecer) muy vivas esperanzas de volver á tenerlo, y yo mismo aldescuido, ó con algun cuidado los he puesto con varios pretextos, en que me dijeron como era el Rey de Kiché, y hablan de /iv/ esto con mucha individualidad. El author de este Kalendario estuvo encarcelado por curandero, pero quebró la carcel, y no se ha visto mas; tengo especie de habérseme dicho que era el maestro de coro.

Añádese aqui lo que dice Fr. Francisco Ximenez en el Capitulo 36 del primer libro de su *Historia de la Provincia de Predicadores de San Vicente de Chiapas y Guatemala* (MS. de la biblioteca de la Universidad de Guatemala, tom. I, foja 87 verso): "De el modo que tenian [los Indios de Guatemala] de contar su año":

Calendar of the Indians of Guatemala, 1722. Kiché. Copied in Guatemala City, April 1877.[1]

Introduction

The original of this calendar in the K'iche' language (erroneously identified as a Kaqchikel calendar in the catalog of the library of the Ethnological Section of the National Museum of Guatemala, 1875, f. 8, no. 1, forms a notebook of twenty-four worthwhile leaves in small quarto. It is found at the beginning of a folio volume titled "Larras, Opusculos."

The second part, beginning on f. 37 of the present copy, is written in another hand in different orthography, bearing in the margin (f. 37 of this copy) the date March 13, 1770.

It appears that it is the same calendar of which the archbishop Don Pedro Cortés y Larraz speaks in his *Descripción Geográfico-Moral de la Diócesis de Guathemala* hecho en la visita que hizo de ella /iii/ en los años de 1768, 1769, y 1770, f. 142v:

> In the town of Quezaltenango I found one of their calendars that I very much wanted to get hold of, and I found another in the town of San Pedro La Laguna,[2] written in the year 1732 in a good hand in the K'iche' language. I have shown them to the best instructors of this language, and although they know the meaning of the terms, none has yet penetrated its importance with the exception of the mast part, in which it is clearly set down enough to know that that day was suitable for having responses sung, celebrating masses, and other devout functions. I understand that among their papers (of the Indians) may be found in the towns rare histories of the king[3] of Quiché, because these Indians have (in my opinion) lively hopes of having kingship again. And I myself have casually and with some care confronted them with various pretexts for telling me about the king of K'iche', and they speak of /iv/ this with great particularity. The author of this calendar was jailed as a curer but broke jail and has not been seen again. I have the feeling that I was told he was choirmaster.

Let us add here what Fray Francisco Ximenez states in chapter 26 of the first volume of his *Historia de la Provincia de Predicadores de San Vicente de Chiapas y Guatemala* (Manuscript of the library of the Universidad de Guatemala, v. 1, f. 87v): "How the Indians of Guatemala counted their year":

No fueron tan bárbaros estos indios como pensaron algunos, que no tuvieron la observancia del movimiento de el Sol para dividir su tiempo; conocieron muy bien y alcanzaron que el año tenia 365 días aunque no alcanzaron la sobra de seis horas, o casí, cada año, por la cual es necesario poner el dia intercalar ó bisiesto. En la division del mes ó semana ó como lo quisieron llamar, iban muy diferentes de /v/ nosotros. El muy R.° P.° M.° Roman en el lib. I Cap. 10 de la República de los Indios dice, que los Mexicanos dividian por meses y estos eran de 20 dias y las semanas de 13 cada una, y que sobraban cinco dias á los cuales llamaban baldios, y en este cómputo entiendo que todos estos Reynos eran conformes, pero señala otros señores ó signos de cada dia, aunque tambien con nombres de animales y otras cosas.

El año de estos empezaba del 21 de Febrero y este era como el dia de año nuevo. Este dia tiene aqueste signo imox, que dice: "envidia del nieto" y hace alocución á la envidia de Hunbatz y Hunchoben á Hunahpu y Xbalanque, porque este nieto es de mujer, no de hombre.

El 2° dia (Febrero 22) tiene el signo ic que significa luna, ó chile.

El 3° dia (Febrero 23) acbal, su significado escaso.

El 4° dia (Febrero 24) cat, en quiché red del maíz, pero dice su significado es /vi/ lagarto.

El 5° dia (Febrero 25) can, amarillo, pero su significado es culebra; debe ser cantí, que es mordedura amarilla.

El 6° dia (Febrero 26) es camoy, toma con el diente ó muerde, era nombre de un Señor del Infierno (Xibalba); su significado dice es la muerte.

El 7° dia (Febrero 27) queh, venado.

El 8° dia (Febrero 28) canel, su significado consejo.

El 9° dia (Marzo 1) toh, nombre de el ídolo. Este significa, paxa pero aqui su significado es aguacero.

El 10° dia (Marzo 2) tzi, perro.

These Indians were not as barbaric as some have thought, assuming that they had no observation of the movement of the sun for marking their time. They knew it very well. They arrived at the fact that the year had 365 days, although they did not attain the excess of six hours, or almost, each year, so that it is necessary to insert the intercalary or leap-year day. In the division of the month or week or whatever you may want to call it, they went very differently from us. The Very Reverend Father Roman in book 1, chapter 10, of the República de los Indios states that the Mexicans divided [the calendar] by months, and these were of twenty days, and weeks, thirteen each one, and there were five days left over that they called "empty," and in this calculation I understand that all these kingdoms were in agreement, but he points out other lords and signs of each day, although also with the names of animals and other things.

Their year begins on February 21 and this was like new year's day. This day has the particular sign Imox, which means "envy of the grandson" and makes reference to the envy of Hunbatz and Hunchoben for Hunahpu and Xbalanque, because this grandson is from the woman and not from the man.

The second day (February 22) has the sign Iq', which means "moon" or "chile."

The third day (February 23) [is] Aq'ab'al, its meaning "scarce."

The fourth day (February 24) [is] K'at; in K'iche', "corn net" but its meaning is /vi/ "lizard."

The fifth day (February 25) [is] Kan, "yellow," but its meaning is "serpent"; it should be *canti*, which is "yellow bite."

The sixth day (February 26) is Kame, "to grab with teeth or bite"; it was the name of the Lord of Hell (Xibalba); its meaning is "death."

The seventh day (February 27) [is] Keej, "deer."

The eighth day (February 28) [is] Q'anil, its meaning "counsel."[4]

The ninth day (March 1) [is] Toj, name of the idol. It means "straw" but here its meaning is "rainstorm."

The tenth day (March 2) [is] Tz'i', "dog."

El 11° dia (Marzo 3) batz, nombre de aquel que se volvió mico, uno fiero con barba de la Verapaz.

El 12° dia (Marzo 4) e, diente.

El 13° dia (Marzo 5) ah, maíz tierno, no asazonado.

El 14° dia (Marzo 6) Balam, tigre.

El 15° dia (Marzo 7) tziquin, pájaro.

/vii/

El 16° dia (Marzo 8) ahmac, pecador; su significado el bicho.

El 17° dia (Marzo 9) noh, que es llenar, y cierta goma; dice su significado es el tenple.

El 18° dia (Marzo 10) tihax, muerde rasgando; su significado dice es el pedernal.

El 19° dia (Marzo 11) cooc, su significado dice que es lluvia.

El 20° dia (Marzo 12) hunahpu, que dicen que bajó al infierno y cumplido aqueste número de 20 dias vuelven a empezar hasta dar vuelta ajustar 360 dias, dando á cada uno el bien ó el mal que á ellos se les antoja, señalando si es buen dia ó malo para el que nació en el y así por aqui bian sus adivinos .

Los otros 5 dias de diferencia llamaban dias cerradas y que no tienen dueño, y cumplido todo el número [de 365 dias] volvian a empezar su año otra vez con el mismo orden. Tambien iban dividiendo de 13 en 13 dias que eran como semanas, mas no señalaban /viii/ fiesta alguna ni dia, como decir lunes ó martes, etc.

Nada dice Ximenez de los meses cuyos nombres resultan del primer calendario en los páginas 18–22 y del segundo en pag. 50.

Con respecto á estos calendarios dice el Obispo Cortés y Larraz en la citada "Descripcion geográfico-moral" f. 225, que uno de los ministros de Quesaltenango le entregaba el calendario:

The eleventh day (March 3) [is] B'atz', name of one who became a howler monkey, a bearded wild amimal of the Verapaz.

The twelfth day (March 4) [is] E, "tooth."

The thirteenth day (March 5) [is] Aj, "tender, unripe corn."

The fourteenth day (March 6) [is] Balam, "tiger."

The fifteenth day (March 7) [is] Tz'ikin, "bird."

The sixteenth day (March 8) [is] Ajmaq, "sinner"; its meaning is "owl."

The seventeenth (March 9) [is] No'j, which means "full" and "a certain resin or incense"; the day name says its meaning is "temple."

The eighteenth day (March 10) [is] Tijax, "biting, scratching"; its meaning is said to be "flint."

The nineteenth day (March 11) [is] Kawoq; its meaning is said to be "rain."

The twentieth day (March 12) [is] Junajpu, who is said to have descended into Hell. And this number of twenty days being fulfilled, they begin over until making a rotation to pass 360 days, giving each one the good or evil that they wish, pointing out whether it is a good or a bad day for someone born on it, and so that way their diviners may see it from here.

The other five days' difference they called closed days without an owner, and completing the entire number (of 365 days) they started their year over again in the same order. They also divided by thirteen to have [groups of] thirteen days that were like weeks, but they did not designate /viii/ any festival at all nor days such as Monday, Tuesday, etc.

Ximenez says nothing about the months, whose names are revealed by ff. 18–22 of the first calendar and f. 50 of the second.

Regarding these calendars, Bishop Cortés y Larraz says in the *Descripcion geográfico-moral* already mentioned (f. 225) that one of the ministers of Quetzaltenango had given him the calendar:

que los Indios tienen para su gobierno, diciendo que este es el Almanak de que se usa en todas las parroquias del Kacchiquel y Kiché, y en el Mam es el mismo, pero escrito en el proprio Idioma. Y me persuado ser él que tendrian en su gentilidad y del que tratan algunos libros.

No me detengo en exponer algunas cosas de dicho Kalendario, que se reduce á dirigir las acciones de los Indios en cada uno de los dias del año, pero acciones buenas, y malas, como con siembras y viages; embriaguezes, idolatrias y deshonestidades /ix/ segun las atribuciones de cada dia que hacen á los animales, y aun al demonio que es á quien suelen dar el mayor culto. De este Kalendario me hablo en la visita de Zamayaᴇ su cura, diciendome que le constaba lo tenian, pero que no podia dar mucha razon de él, porque lo ocultaban mucho.

El sugeto que tiene alguna intelegencia del Kalendario de los Indios, que es Fr. Thomas Arrevillaga de la Orden de San Francisco vino accompañandome desde Quesaltenango y dijo al cura de Totonicapan: hoy es dia muy favorable en el Kalendario de los Indios; habra tenido Vmd. muchas misas, responsos y ofrendas. Respondió el cura: es cierto que ha habido de estas cosas con mas abundancia que otros dias . . .

Y en foja 227:

En este Pueblo (S. Pedro de la Laguna) se halló otro Kalendario, que para en mi poder; dijo el cura, que no lo entiende y que los que los forman son los mismos que son curanderos supersticiosos; que es casí imposíble el descubrirlas por el /x/ secreto sumo, que en esto se guarda - - que se pregunta frecuentemente en la confesion por este pecado, pero que todos lo niegan.

/xi/

that the Indians have for their governance, saying that this was their almanac that is in use in all the parishes of Kaqchikel and K'iche', and in Mam it is the same but written in their own language. And I am pursuaded that it is the one they must have had in their pagan days, which is treated in some books.

I shall not pause to expound some things about this calendar, which comes down to directing the actions of the Indians on each one of the days of the year, only good and bad actions, such as planting and travels, drinking bouts, idolatry and immorality, /ix/ according to the attributes ascribed to the animals and even the Devil, who is the one to whom they give the greatest worship. It was of this calendar that the preaching curate of Samayac spoke to me on my visit, telling me that he knew they had it but he could not make much sense out of it because they kept it quite hidden.

One who has some understanding of the calendar of the Indians, Fray Thomas Arrevillaga of the Order of San Francisco, was accompanying me from Quetzaltenango together with a curate of Totonicapán, and he said: Today is a very favorable day in the Indians' calendar. Your Grace will be seeing many masses, responses and devout offices. The curate responded, It is true that there have been those things more abundantly than on other days.

And on folio 227:

In this town (S. Pedro de la Laguna) another calendar was found, which is in my possession, said the curate, but I do not understand it. The ones who make them are the same ones who are superstitious curers; it is almost impossible to find them because of the /x/ great secrecy they keep about this. This [sin] is asked about very frequently in confession, but they all deny it.

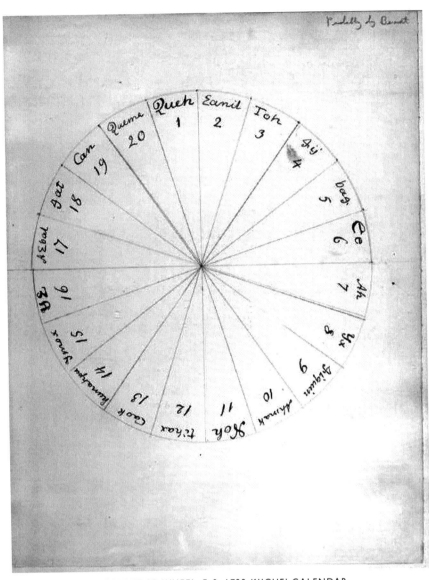

3.1. DRAWING OF CALENDAR WHEEL, F. 3, 1722 K'ICHE' CALENDAR.

Calendar A

/1/

Chol Poal Q'ij	Count Cycle of Days
Vae chol poal εih	This is the count (cycle) of days,
maceval εih,	the macehual days,[5]
co(h)cha chi rech:	as we call them.
vae cahauar chupam hun hunab	This one reigns in one year.
vae nabe mixchap chupam rahauarem.	This is the first [day] that was just taken in its reign.
Noh, Yε, Queh, E.	No'j,[6] Iq',[7] Keej,[8] E.[9]

Xaki e cahib caquihal quib	Only these four replace each other
chupam ri huhun chuhunab.	in each year.
Roo ri hun ri cu4am rahauarem;	The fifth is the one who takes his lordship;
que(he) va cubih	like this they are named:
4 Yε, 5 Aεbal, 6 4at, 7 Cana, 8 Queme.	4 Iq', 5 Aq'ab'al,[10] 6 K'at,[11] 7 Kan,[12] 8 Kame.[13]

*9 queh 4ut mixeka(n) 20 εih nabeme(n)	d	mayo 3	*9 Keej then has just taken the burden[14] [on] the 20th[15] day of Nab'e Mam[16]
10 εanil	e	4	10 Q'anil[17]
11 toh	f	5	11 Toj[18]
12 4,ij	g	6	12 Tz'i'[19]
*13 ba4	A	7	*13 B'atz'[20]
1 ee	b	8	1 E
2 ah	c	9	2 Aj[21]
3 yx	d	10	3 I'x[22]
4 4,iquin	e	11	4 Tz'ikin[23]
5 ahmak	f	12	5 Ajmaq[24]

6 noh	g	13	6 No'j
7 tihax	A	14	7 Tijax[25]

/2/

8 caok	b	mayo 15	8 Kawoq[26]
9 hunahpu	c	16	9 Junajpu[27, 28]
10 ymox	d	17	10 Imox[29]
11 yɛ	e	18	11 Iq'
12 aɛbal	f	19	12 Aq'ab'al
*13 4at	g	20	*13 K'at
1 can	A	21	1 Kan
2 queme	b	22	2 Kame
*3 queh mixeka(n) 20 ɛih vcab mam	c	23	3 Keej has just taken the burden [on] the 20th day of Ukab' Mam[30]
4 ɛanil	d	24	4 Q'anil
5 toh	e	25	5 Toj
6 4,ij	f	26	6 Tz'i'
7 ba4	g	27	7 B'atz'
8 ee	A	28	8 E
9 ah	b	29	9 Aj
10 yx	c	30	10 I'x
11 4,iquin	d	31	11 Tz'ikin
12 ahmak	e	junio 1	12 Ajmaq
*13 noh	f	2	*13 No'j
1 tihax	g	3	1 Tijax
2 caok	A	4	2 Kawoq
3 hunahpu	b	5	3 Junajpu

/3/

4 ymox	c	junio 6	4 Imox

5 yε	d	7	5 Iq'
6 aεbal	e	8	6 Aq'ab'al
7 4at	f	9	7 K'at
8 can	g	10	8 Kan
9 queme	A	11	9 Kame
*10 queh mixeka(n) chi 20 εih liqinca	b	12	*10 Keej has just taken the burden on the 20th day of Likinka[31]
11 εanil	c	13	11 Q'anil
12 toh	d	14	12 Toj
*13 4,ij	e	15	*13 Tz'i'
1 ba4	f	16	1 B'atz'
2 ee	g	17	2 E
3 ah	A	18	3 Aj
4 yx	b	19	4 I'x
5 4,iquin	c	20	5 Tz'ikin
6 ahmak	d	21	6 Ajmaq
7 noh	e	22	7 No'j
8 tihax	f	23	8 Tijax
9 caok	g	24	9 Kawoq
10 hunahpu	A	25	10 Junajpu
11 ymox	b	26	11 Imox
12 yε	c	27	12 Iq'
*13 εkbal	d	28	*13 Aq'ab'al

/4/

1 4at	e	junio 29	1 K'at
2 can	f	30	2 Kan
3 queme	g	julio 1	3 Kame
*4 queh mixekan chi 20 εih vcab liquinca	A	2	*4 Keej has just taken the burden on the 20th day of Ukab' Likinka

5 ɛanil	b	3	5 Q'anil
6 toh	c	4	6 Toj
7 4,ij	d	5	7 Tz'i'
8 ba4	e	6	8 B'atz'
9 ee	f	7	9 E
10 ah	g	8	10 Aj
11 yx	A	9	11 I'x
12 4,iquin	b	10	12 Tz'ikin
*13 ahmak	c	11	*13 Ajmaq
1 noh	d	12	1 No'j
2 tihax	e	13	2 Tijax
3 caok	f	14	3 Kawoq
4 hunahpu	g	15	4 Junajpu
5 ymox	A	16	5 Imox
6 yɛ	b	17	6 Iq'
7 aɛbal	c	18	7 Aq'ab'al
8 4at	d	19	8 K'at
9 can	e	20	9 Kan

/5/

10 queme	f	julio 21	10 Kame
*11 queh mixɛkan chi 20 ɛih nab pach	g	22	*11 Keej has just taken the burden on the 20th day of Nab'e Pach[32]
12 ɛanil	A	23	12 Q'anil
*13 toh	b	24	*13 Toj
1 4,ij	c	25	1 Tz'i'
2 ba4	d	26	2 B'atz'
3 ee	e	27	3 E
4 ah	f	28	4 Aj
5 yx	g	29	5 I'x
6 4,iquin	A	30	6 Tz'ikin

7 ahmak	b	31	7 Ajmaq
8 noh	c	agosto 1	8 No'j
9 tihax	d	2	9 Tijax
10 caok	e	3	10 Kawoq
11 hunahpu	f	4	11 Junajpu
12 ymox[,] nima chaomalah ɛih	g	5	12 Imox. A very beautiful day
*13 yɛ	A	6	*13 Iq'
1 aɛbal	b	7	1 Aq'ab'al
2 4at	c	8	2 K'at
3 can	d	9	3 Kan
4 queme	e	10	4 Kame
*5 queh mixekan chi 20 ɛih vcab pach	f	11	*5 Keej has just taken the burden on the 20th day of Ukab' Pach.

/6/

6 ɛanil	g	agosto 12	6 Q'anil
7 toh	A	13	7 Toj
8 4,ij	b	14	8 Tz'i'
9 ba4	c	15	9 B'atz'
10 ee	d	16	10 E
11 ah	e	17	11 Aj
12 yx	f	18	12 I'x
*13 4,iquin	g	19	*13 Tz'ikin
1 ahmak	A	20	1 Ajmaq
2 noh	b	21	2 No'j
3 tihax	c	22	3 Tijax
4 caok	d	23	4 Kawoq
5 hunahpu	e	24	5 Junajpu
6 ymox	f	25	6 Imox
7 yɛ	g	26	7 Iq'

8 aɛbal	A	27	8 Aq'ab'al
9 4at	b	28	9 K'at
10 can	c	29	10 Kan
11 queme	d	30	11 Kame
*12 queh mixeka(n) chi 20 ɛih 4,icilakam	e	31	*12 Keej has just taken the burden on the 20th day of Tz'isi Laqam[33]
*13 ɛanil	f	septiembre 1	*13 Q'anil
1 toh	g	2	1 Toj

/7/

2 4,ij	A	septiembre 3	2 Tz'i'
3 ba4	b	4	3 B'atz'
4 ee	c	5	4 E
5 ah	d	6	5 Aj
6 yx	e	7	6 I'x
7 4,iquin	f	8	7 Tz'ikin
8 ahmak	g	9	8 Ajmaq
9 noh	A	10	9 No'j
10 tihax	b	11	10 Tijax
11 caok	c	12	11 Kawoq
12 hunahpu	d	13	12 Junajpu
*13 ymox	e	14	*13 Imox
1 yɛ	f	15	1 Iq'
2 aɛbal	g	16	2 Aq'ab'al
3 4at	A	17	3 K'at
4 can	b	18	4 Kan
5 queme	c	19	5 Kame
*6 queh mixkube chi 20 ɛih 4,iquin ɛih	d	20	*6 Keej has just taken office[34] on the 20th day of Tz'ikin Q'ij[35]

7 ɛanil	e	21	7 Q'anil
8 toh	f	22	8 Toj
9 4,ij	g	23	9 Tz'i'
10 ba4	A	24	10 B'atz'
11 ee	b	25	11 E

/8/

12 ah	c	septiembre 26	12 Aj
*13 yx	d	27	*13 I'x
1 4,iquin	e	28	1 Tz'ikin
2 ahmak	f	29	2 Ajmaq
3 noh	g	30	3 No'j
4 tihax	A	octubre 1	4 Tijax
5 caok	b	2	5 Kawoq
6 hunahpu	c	3	6 Junajpu
7 ymox	d	4	7 Imox
8 yɛ	e	5	8 Iq'
9 aɛbal	f	6	9 Aq'ab'al
10 4at	g	7	10 K'at
11 can	A	8	11 Kan
12 queme	b	9	12 Kame
**13 queh mixeka(n) chi 20 ɛih cakam	c	10	**13 Keej has just taken the burden on the 20th day of Kaqam[36]
1 ɛanil	d	11	1 Q'anil
2 toh	e	12	2 Toj
3 4,ij	f	13	3 Tz'i'
4 ba4	g	14	4 B'atz'
5 ee	A	15	5 E
6 ah	b	16	6 Aj
7 yx	c	17	7 I'x

/9/

8 4,iquin	d	octubre 18	8 Tz'ikin
9 ahmak	e	19	9 Ajmaq
10 noh	f	20	10 No'j
11 tihax	g	21	11 Tijax
12 caok	A	22	12 Kawoq
*13 hunahpu	b	23	*13 Junajpu
1 ymox	c	24	1 Imox
2 yɛ	d	25	2 Iq'
3 aɛbal	e	26	3 Aq'ab'al
4 4at	f	27	4 K'at
5 can	g	28	5 Kan
6 queme	A	29	6 Kame
*7 queh mixeka(n) chi 20 ɛih botam	b	30	*7 Keej has just taken the burden on the 20th day of B'otam[37]
8 ɛanil	c	31	8 Q'anil
9 toh	d	noviembre 1	9 Toj
10 4,ij	e	2	10 Tz'i'
11 ba4,	f	3	11 B'atz'
12 ee	g	4	12 E
*13 ah	A	5	*13 Aj
1 yx	b	6	1 I'x
2 4,iquin	c	7	2 Tz'ikin
3 ahmak	d	8	3 Ajmaq

/10/

4 noh	e	noviembre 9	4 No'j
5 tihax	f	10	5 Tijax
6 caok	g	11	6 Kawoq
7 hunahpu	A	12	7 Junajpu

8 ymox	b	13	8 Imox
9 yε	c	14	9 Iq'
10 aεbal	d	15	10 Aq'ab'al
11 4at	e	16	11 K'at
12 can	f	17	12 Kan
*13 queme[,] couilah εih	g	18	*13 Kame; a very powerful day
*1 queh mixekan chi 20 εih nabe εih	A	19	*1 Keej has just taken the burden on the 20th day of Nab'e Si'j[38]
2 εanil	b	20	2 Q'anil
3 toh	c	21	3 Toj
4 4,ij	d	22	4 Tz'i'
5 ba4	e	23	5 B'atz'
6 ee	f	24	6 E
7 ah	g	25	7 Aj
8 yx	A	26	8 I'x
9 4,iquin	b	27	9 Tz'ikin
10 ahmak	c	28	10 Ajmaq
11 noh	d	29	11 No'j
12 tihax	e	30	12 Tijax

/11/

*13 caok	f	diciembre 1	*13 Kawoq
1 hunahpu	g	2	1 Junajpu
2 ymox	A	3	2 Imox
3 yε	b	4	3 Iq'
4 aεbal	c	5	4 Aq'ab'al
5 4at	d	6	5 K'at
6 can	e	7	6 Kan
7 queme	f	8	7 Kame

*8 queh mixeka(n) chi 20 ɛih vcab çih	g	9	*8 Keej has just taken the burden on the 20th day of Ukab' Si'j
9 ɛanil	A	10	9 Q'anil
10 toh	b	11	10 Toj
11 4,ij	c	12	11 Tz'i'
12 ba4	d	13	12 B'atz'
*13 ee	e	14	*13 E
1 ah	f	15	1 Aj
2 yx	g	16	2 I'x
3 4,iquin	A	17	3 Tz'ikin
4 ahmak	b	18	4 Ajmaq
5 noh	c	19	5 No'j
6 tihax	d	20	6 Tijax
7 caok	e	21	7 Kawoq
8 hunahpu	f	22	8 Junajpu
9 ymox	g	23	9 Imox

/12/

10 yɛ	A	diciembre 24	10 Iq'
11 aɛbal	b	25	11 Aq'ab'al
12 4at	c	26	12 K'at
*13 can	d	27	*13 Kan
1 queme	e	28	1 Kame
2 queh mixekan chi 20 rox çih	f	29	*2 Keej has just taken the burden on the 20th day of Rox Si'j
3 ɛanil	g	30	3 Q'anil
4 toh	A	31	4 Toj
5 4,ij	b	enero 1	5 Tz'i'
6 ba4	c	2	6 B'atz'
7 ee	d	3	7 E

8 ah	e	4	8 Aj
9 yx	f	5	9 I'x
10 4,iquin	g	6	10 Tz'ikin
11 ahmak	A	7	11 Ajmaq
12 noh	b	8	12 No'j
*13 tihax	c	9	*13 Tijax
1 caok	d	10	1 Kawoq
2 hunahpu	e	11	2 Junajpu
3 ymox	f	12	3 Imox
4 yɛ	g	13	4 Iq'
5 aɛbal	A	14	5 Aq'ab'al

/13/

6 4at	b	15	6 K'at
7 can	c	16	7 Kan
8 queme	d	17	8 Kame
9 queh mixeka[n] chi 20 ɛih chee	e	18	*9 Keej has just taken the burden on the 20th day of Che'[39]
10 ɛanil	f	19	10 Q'anil
11 toh	g	20	11 Toj
12 4,ij	A	21	12 Tz'i'
*13 ba4	b	22	*13 B'atz'
1 ee	c	23	1 E
2 ah	d	24	2 Aj
3 yx	e	25	3 I'x
4 4,iquin	f	26	4 Tz'ikin
5 ahmak[,] varal caticar avex ma tak huyub	g	27	5 Ajmaq, here begins the sowing of milpa [but] not in the mountains
6 noh	A	28	6 No'j

7 tihax	b	29	7 Tijax
8 caok	c	30	8 Kawoq
9 hunahpu	d	31	9 Junajpu
10 ymox	e	febrero 1	10 Imox
11 yɛ	f	2	11 Iq'
12 aɛbal	g	3	12 Aq'ab'al
*13 4at	A	4	*13 K'at
1 can	b	5	1 Kan

/14/

2 queme	c	febrero 6	2 Kame
3 queh mixekan chi 20 ɛih tequexepual[;] are cavex vi ronohel vinak	d	7	*3 Keej has just taken the burden on the 20th day of Tekexepoal;[40] [this is the time] when all people plant
4 ɛanil	e	8	4 Q'anil
5 toh	f	9	5 Toj
6 4,ij	g	10	6 Tz'i'
7 ba4	A	11	7 B'atz'
8 ee	b	12	8 E
9 ah	c	13	9 Aj
10 yx	d	14	10 I'x
11 4,iquin	e	15	11 Tz'ikin
12 ahmak	f	16	12 Ajmaq
*13 noh	g	17	*13 No'j
1 tihax	A	18	1 Tijax
2 caok	b	19	2 Kawoq
3 hunahpu	c	20	3 Junajpu
4 ymox	d	21	4 Imox
5 yɛ	e	22	5 Iq'
6 aɛbal	f	23	6 Aq'ab'al

7 4at	g	24	7 K'at
8 can	A	25	8 Kan
9 queme	b	26	9 Kame
*10 <u>queh</u> mixekan chi 20 ɛih <u>4,ibapopp</u>	c	27	*10 <u>Keej</u> has just taken the burden on the 20th day of <u>Tz'ib'a Pop</u>[41]
11 ɛanil	d	28	11 Q'anil

/15/

12 toh	e	março 1	12 Toj
*<u>13</u> 4,ij	f	2	*<u>13</u> Tz'i'
1 ba4	g	3	1 B'atz'
2 ee	A	4	2 E
3 ah	b	5	3 Aj
4 yx	c	6	4 I'x
5 4,iquin	d	7	5 Tz'ikin
6 ahmak	e	8	6 Ajmaq
7 noh	f	9	7 No'j
8 tihax	g	10	8 Tijax
9 caok	A	11	9 Kawoq
10 hunahpu	b	12	10 Junajpu
11 ymox	c	13	11 Imox
12 yɛ	d	14	12 Iq'
*<u>13</u> aɛbal	e	15	*<u>13</u> Aq'ab'al
1 4at	f	16	1 K'at
2 can	g	17	2 Kan
3 queme 4isbal avex	A	18	3 Kame; end of planting
4 <u>queh</u> mixekan chi 20 ɛih <u>çac</u>	b	19	*4 <u>Keej</u> has just taken the burden on the 20th day of <u>Saq</u>[42]
5 ɛanil	c	20	5 Q'anil

6 toh	d	21	6 Toj
7 4,ij	e	22	7 Tz'i'
8 ba4	f	23	8 B'atz'

/16/

9 ee	g	março 24	9 E
10 ah	A	25	10 Aj
11 yx	b	26	11 I'x
12 4,iquin	c	27	12 Tz'ikin
*13 ahmak nima tabal ɛih	d	28	*13 Ajmaq; an important petition[43]
1 noh	e	29	1 No'j
2 tihax	f	30	2 Tijax
3 caok	g	31	3 Kawoq
4 hunahpu	A	abril 1	4 Junajpu
5 ymox	b	2	5 Imox
6 yɛ	c	3	6 Iq'
7 aɛbal	d	4	7 Aq'ab'al
8 4at	e	5	8 K'at
9 can	f	6	9 Kan
10 queme	g	7	10 Kame
*11 queh mixekan chi 20 ɛih 4hab nima roquibal yaoh	A	8	*11 Keej has just taken the burden on the 20th day of Ch'ab';[44] great entering of offerings
12 ɛanil	b	9	12 Q'anil
*13 toh	c	10	*13 Toj
1 4,ij	d	11	1 Tz'i'
2 ba4	e	12	2 B'atz'
3 ee	f	13	3 E

/17/

4 ah	g	abril 14	4 Aj
5 yx	A	15	5 I'x
6 4,iquin	b	16	6 Tz'ikin
7 ahmak	c	17	7 Ajmaq
8 noh	d	18	8 No'j
9 tihax	e	19	9 Tijax
10 caok	f	20	10 Kawoq
11 hunahpu	g	21	11 Junajpu
12 ymox	A	22	12 Imox
*13 yε	b	23	*13 Iq'
1 aεbal	c	24	1 Aq'ab'al
2 4at	d	25	2 K'at
3 can	e	26	3 Kan
4 queme	f	27	4 Kame
*5 queh mix4iz chi 20 εih 4izbal rech	g	28	*5 Keej has just finished on the 20th day of K'isb'al Rech[45]
6 εanil	A	29	6 Q'anil
7 toh	b	30	7 Toj

Va 4ut hoob εih	And these are the five [days],
4api εih mahi rahauh	closing days, [46] without lord,
cohchaa chirech	as we call it,
quehe va cubih	like this it is named.
6 εanil mixutic vi uloε	6 Q'anil has just planted it hither.

8 4,ij	c	mayo 1	8 Tz'i'
8 ba4	d	2	9 B'atz'

/18/

10 ee	e	3	10 E
11 ah	f	4	11 Aj

Vae 20 chi ɛih 10 ee nabemam	This is the 20th on the day 10 E Nab'e Mam,
cahauan hun hunab	he reigns one year.
Noh, Yɛ, Queh, Ee. 1723 años.	No'j, Iq', Keej, E. Year 1723.
10 e mix[hue] 20 nabemá chupam 4 Mayo xu4amo	10 E has just sustained[47] the 20th [day] of Nab'e Mam on May 4.[48] He received them.
4 e mixu4am chi 20 vcab má 23 Mayo	4 E has just sustained on the 20th [day] of Ukab' Mam on May 23.
11 e mixu4am chi 20 liquinca 12 Junio xu4amo	11 E has just sustained on the 20th [day] of Likinka on June 12. He received them.
5 e mixu4am chi 20 vcab liquinca 2 Julio	5 E has just sustained on the 20th [day] of Ukab' Likinka on July 2.
12 e mixekan chi 20 nabe pach 22 Julio	12 E has just taken the burden on the 20th [day] of Nab'e Pach on July 22.
6 e mixekan chi 20 ucab pach 12 Agosto	6 E has just taken the burden on the 20th [day] of Ukab' Pach on August 12.[49]
13 e mixekan chi 20 4,içilakam 1 Septiembre	*13 E has just taken the burden on the 20th [day] of Tz'isi Laqam on September 1.[50]
7 e mixekan chi 20 4,iquin ɛih 12 Septiembre	7 E has just taken the burden on the 20th [day] of Tz'ikin Q'ij on September 12.[51]
1 e mixekan chi 20 cakam 12 Octubre	1 E has just taken the burden on the 20th [day] of Kaqam on October 12.[52]
8 e mixekan chi 20 botam 1 Noviembre	8 E has just taken the burden on the 20th [day] of B'otam on November 1.[53]
2 e mixekan chi 20 nabeçih 19 Noviembre	2 E has just taken the burden on the 20th [day] of Nab'e Si'j on November 19.
9 e mixekan chi 20 vcabçih 9 Diciembre	9 E has just taken the burden on the 20th [day] of Ukab' Si'j on December 9.
3 e mixekan chi 20 roxçih 29 Diciembre	3 E has just taken the burden on the 20th [day] of Rox Si'j on December 29.

10 e mixekan chi 20 chee 18 Enero	10 E has just taken the burden on the 20th [day] of Che' on January 18.
4 e mixekan chi 20 tequexepual 7 Febrero	4 E has just taken the burden on the 20th [day] of Tekexepoal on February 7.
11 e mixekan chi 20 4,ibap[o]p 27 Febrero	11 E has just taken the burden on the 20th [day] of Tz'ib'a Pop on February 27.
5 e mixekan chi 20 çac 20 Março	5 E has just taken the burden on the 20th [day] of Saq on March 20.[54]
12 e mixekan chi 20 4hab 9 Abril	12 E has just taken the burden on the 20th [day] of Ch'ab' on April 9.[55]
6 e mixu4iso	6 E has just finished it.
vae 4ut hoob 4apiɛihih mahi raha	And these are the five closing days without lord,

/19/

uh coh chaa chirech.	as we call them,
7 ah, 8 yx, 9 4,iquin, 10 ahmak, 11 noh	7 Aj, 8 I'x, 9 Tz'ikin, 10 Ajmaq, 11 No'j,
xu4am rahauarem.	has taken his reign.
Noh, Yɛ, Queh, E. 1724 años.	No'j, Iq', Keej, E. Year 1724.
11 noh mixu4am 20 ɛih nabemam	11 No'j has just sustained the 20th day of Nab'e Mam.
5 noh xu4an chi 20 vcab mam	5 No'j sustained on the 20th day of Ukab' Mam.
12 noh xu4an chi 20 nabe liquinca	12 No'j sustained on the 20th day of Nab'e Likinka.
6 noh xu4an chi 20 vcab liquinca	6 No'j sustained on the 20th day of Ukab' Likinka.
13 noh xu4an chi 20 nabe pach	13 No'j sustained on the 20th day of Nab'e Pach.
7 noh xu4an chi 20 vcab pach	7 No'j sustained on the 20th day of Ukab' Pach.
1 noh xu4an chi 20 4içi lakam	1 No'j sustained on the 20th day of Tz'isi Laqam.

8 noh xu4an chi 20 4,iquin εih	8 No'j sustained on the 20th day of Tz'ikin.
2 noh xu4an chi 20 cakam	2 No'j sustained on the 20th day of Kaqam.
9 noh xu4an chi 20 botam	9 No'j sustained on the 20th day of B'otam.
3 noh xu4an chi 20 nabeçih	3 No'j sustained on the 20th day of Nab'e Si'j.
10 noh xu4an chi 20 vcabecih	10 No'j sustained on the 20th day of Ukab' Si'j.
4 noh xu4an chi 20 rox çih	4 No'j sustained on the 20th day of Rox Si'j.
11 noh xu4an chi 20 chee	11 No'j sustained on the 20th day of Che'.
5 noh xu4an chi 20 tequexepual	5 No'j sustained on the 20th day of Tekexepoal.
12 noh xu4an chi 20 4iba popp	12 No'j sustained on the 20th day of Tz'ib'a Pop.
6 noh xu4an chi 20 çac	6 No'j sustained on the 20th day of Saq.
13 noh xu4an chi 20 4hab	13 No'j sustained on the 20th day of Ch'ab'.
7 vucub noh mix4iz cu	7 Seven No'j has just finished then.
vae 4ut hoob 4apiεih	And these are the five closing days,

/20/

4apiεih,	closing days,
quehe	like:
vucub 7 noh 8 tihax 9 caok 10 hunahpu 11 ymox 12 yε	7 No'j, 8 Tijax, 9 Kawoq, 10 Junajpu, 11 Imox, 12 Iq',
4ut	then.

Noh Yε Queh Ee 1725 años.	No'j. Iq'. Keej. E. Year 1725.
12 yε mixekan chi 20 εih nabe mam	12 Iq' has just taken the burden on the 20th day of Nab'e Mam.
6 yε mixu4an chi 20 εih vcab mam	6 Iq' has just sustained on the 20th day of Ukab' Mam.
13 yε mixu4an chi 20 εih nabe liquin ca	13 Iq' has just sustained on the 20th day of Nab'e Likinka.

7 yɛ mixu4an chi 20 ɛih vcab liquinca	7 Iq' has just sustained on the 20th day of Ukab' Likinka.
1 yɛ mixu4an chi 20 ɛih nabe pach	1 Iq' has just sustained on the 20th day of Nab'e Pach.
8 yɛ mixu4an chi 20 ɛih vcab pach	8 Iq' has just sustained on the 20th day of Ukab' Pach.
2 yɛ mixu4an chi 20 ɛih 4,içi lakam	2 Iq' has just sustained on the 20th day of Tz'isi Laqam.
9 yɛ mixu4an chi 20 ɛih 4,iquin ɛih	9 Iq' has just sustained on the 20th day of Tz'ikin Q'ij.
3 yɛ mixu4an chi 20 ɛih cakam	3 Iq' has just sustained on the 20th day of Kaqam.
10 yɛ mixu4an chi 20 ɛih botam	10 Iq' has just sustained on the 20th day of B'otam.
4 yɛ mixu4an chi 20 ɛih nabeçih	4 Iq' has just sustained on the 20th day of Nab'e Si'j.
11 yɛ mixu4an chi 20 ɛih vcabçih	11 Iq' has just sustained on the 20th day of Ukab' Si'j.
5 yɛ mixu4an chi 20 ɛih roxçih	5 Iq' has just sustained on the 20th day of Rox Si'j.
12 yɛ mixu4an chi 20 ɛih chee nabe avabal 13 Juno	12 Iq' has just sustained on the 20th day of Che'; first place for sowing on June 13.
6 yɛ mixu4an chi 20 ɛih tequexepual ronohel cavaxic	6 Iq' has just sustained on the 20th day of Tekexepoal; everything is sown.
13 yɛ mixu4an chi 20 ɛih 4iba popp	13 Iq' has just sustained on the 20th day of Tz'ib'a Pop.
7 yɛ mixu4an chi 20 ɛih çac	7 Iq' has just sustained on the 20th day of Saq.
1 yɛ mixu4an chi 20 ɛih 4hab	1 Iq' has just sustained on the 20th day of Ch'ab'.
8 yɛ mix4iz rahauarem vae yɛ	8 Iq' has just finished the reign of this Iq'.
vae 4ut hoob 4api ɛih	And these are the five closing days,
quehe va cubih	like this they are named,
8 íɛ 9 aɛbal 10 4at 11 can 12 queme	8 Iq', 9 Aq'ab'al, 10 K'at, 11 Kan, 12 Kame,

| 13 quehe 4ut nabemam | then 13 Keej Nab'e Mam. |

/21/

Noh Yɛ Queh Ee 1726 años.	No'j. Iq'. Keej. E. Year 1726.
13 queh mixekan 20 ɛih nabe mam	13 Keej has just taken the burden [on] the 20th day of Nab'e Mam.
7 queh mixu4an chi 20 ɛih vcab mam	7 Keej has just sustained on the 20th day of Ukab' Mam.
1 queh mixu4an chi 20 ɛih nabe liquin ca	1 Keej has just sustained on the 20th day of Nab'e Likinka.
8 queh mixu4an chi 20 ɛih vcab liquinca	8 Keej has just sustained on the 20th day of Ukab' Likinka.
2 queh mixu4an chi 20 ɛih nabe pach	2 Keej has just sustained on the 20th day of Nab'e Pach.
9 queh mixu4an chi 20 ɛih vcab pach	9 Keej has just sustained on the 20th day of Ukab' Pach.
3 queh mixu4an chi 20 ɛih 4,içi lakam	3 Keej has just sustained on the 20th day of Tz'isi Laqam.
10 queh mixu4an chi 20 ɛih 4,iquin ɛih	10 Keej has just sustained on the 20th day of Tz'ikin Q'ij.
4 queh mixu4an chi 20 ɛih cakam	4 Keej has just sustained on the 20th day of Kaqam.
11 queh mixu4an chi 20 ɛih botam	11 Keej has just sustained on the 20th day of B'otam.
5 queh mixu4an chi 20 ɛih nabe çih	5 Keej has just sustained on the 20th day of Nab'e Si'j.
12 queh mixu4an chi 20 ɛih vcab çih	12 Keej has just sustained on the 20th day of Ukab' Si'j.
6 queh mixu4an chi 20 ɛih rox çih	6 Keej has just sustained on the 20th day of Rox Si'j.
13 queh mixu4an chi 20 ɛih chee	13 Keej has just sustained on the 20th day of Che'.
7 queh mixu4an chi 20 ɛih tequexepual	7 Keej has just sustained on the 20th day of Tekexepoal.

1 queh mixu4an chi 20 ɛih 4ihba popp	1 Keej has just sustained on the 20th day of Tz'ib'a Pop.
8 queh mixu4an chi 20 ɛih çac	8 Keej has just sustained on the 20th day of Saq.
2 queh mixu4an chi 20 ɛih 4hab	2 Keej has just sustained on the 20th day of Ch'ab'.
9 queh mi[x]u4is rahavarem vae queh	9 Keej has just finished the reign of this Keej.
vae 4ute hoob 4,apiɛih	And these are the five closing days,
vae cubih	[like] this they are named:
10 ɛanil 11 toh 12 4,ij 13 ba4	10 Q'anil, 1 Toj, 1 Tz'i', 13 B'atz',
1 ee 4ut	and then 1 E
ru4am rahavarem chupam hun chi hunab.	sustained the reign in another year.

/22/

Noh Yɛ Queh E 1727 años	No'j. Iq'. Keej. E. Year 1727.
1 E mixekan 20 ɛih nabe mam	1 E has just taken the burden [on] the 20th day of Nab'e Mam.
8 e mixu4an chi 20 ɛih vcab mam	8 E has just sustained on the 20th day of Ukab' Mam.
2 e mixu4an chi 20 ɛih nabe liquin ca	2 E has just sustained on the 20th day of Nab'e Likinka.
9 e mixu4an chi 20 ɛih vcab liquin ca	9 E has just sustained on the 20th day of Ukab' Likinka.
3 e mixu4an chi 20 ɛih nabe pach	3 E has just sustained on the 20th day of Nab'e Pach.
10 e mixu4an chi 20 ɛih vcab pach	10 E has just sustained on the 20th day of Ukab' Pach.
4 e mixu4an chi 20 ɛih 4,içi lakam	4 E has just sustained on the 20th day of Tz'isi Laqam.
11 e mixu4an chi 20 ɛih 4,iquin ɛih	11 E has just sustained on the 20th day of Tz'ikin Q'ij.
5 e mixu4an chi 20 ɛih cakam	5 E has just sustained on the 20th day of Kaqam.

12 e mixu4an chi 20 ɛih botam	12 E has just sustained on the 20th day of B'otam.
6 e mixu4an chi 20 ɛih nabe çih	6 E has just sustained on the 20th day of Nab'e Si'j.
13 e mixu4an chi 20 ɛih vcab çih	13 E has just sustained on the 20th day of Ukab' Si'j.
7 e mixu4an chi 20 ɛih rox çih	7 E has just sustained on the 20th day of Rox Si'j.
1 e mixu4an chi 20 ɛih chee	1 E has just sustained on the 20th day of Che'.
8 e mixu4an chi 20 ɛih tequexepual	8 E has just sustained on the 20th day of Tekexepoal.
2 e mixu4an chi 20 ɛih 4iba popp	2 E has just sustained on the 20th day of Tz'ib'a Pop.
9 e mixu4an chi 20 ɛih çac	9 E has just sustained on the 20th day of Saq.
3 e mixu4an chi 20 ɛih 4hab	3 E has just sustained on the 20th day of Ch'ab'.
10 E mixu4is rahavarem vae e	10 E has just finished the reign of this E.
vae 4ute hoob çapi ɛih	And these are the five closing days,
quehe va cubih	like this they are named:
11 ah 12 yx 13 4,iquin 1 ahmak 2 noh 4ut	11 Aj, 12 I'x, 13 Tz'ikin, 1 Ajmaq, and then 2 No'j.

[Calendar B]
[Divinatory Calendar 1]

/23/

Ahilabal ɛih		*Count of the Day*	
1 ymox	[T]icbal ɛih	1 Imox	Planting day,
1 ah	avexabal ɛih	1 Aj	sowing day,
1 can	vtzilah ɛih	1 Kan	a very good day,

1 noh	roo ychal	1 No'j	all five.[56]
1 thoh		1 Toj	
2 yɛ	Ahcakvach	2 Iq'	He of hate,[57]
2 yix	ahmoxvach	2 I'x	he of jealousy,[58]
2 queme	ahqueeba4ux	2 Kame	the one of two hearts,[59]
2 tihax	c[h]i nah vnaoh	2 Tijax	far away [is] his mind,[60]
2 4,ij	katzih ytzel ɛih	2 Tz'i'	a truly evil day,
	roo ychal		all five.
3 akbal	Ytzel ɛih	3 Aq'ab'al	An evil day,
3 4,iquin	4a chi nah catzihon vi	3 Tz'ikin	from far away speaks
3 queh	v4ux vnnaoh	3 Keej	his heart, his mind,[61]
3 caok	roo yichal	3 Kawoq	all five,
3 ba4,	xax ma vtz ta ui	3 B'atz'	extremely bad is
	rij chi4uxlaax		what shall be remembered
	chupam.		within [these five days].
4 4at	Ma na vtzilah ɛih tah	4 K'at	Not a very good day,
4 ahmak	ma na ka tzih ta	4 Ajmaq	truly not,
4 ɛanil	cacouinic	4 Q'anil	it is becoming powerful
4 hunahpu	chi ɛih	4 Junajpu	each day,
4 ee	katzil	4 E	truly,
	nabe royoval ɛih		first day of anger,
	roo ychal.		all five.

/24/

5 can	Hun achij hun yxok	5 Kan	A man, a woman,
5 noh	ytzel vach	5 No'j	of bad character,
5 thoh	quemacunic	5 Toj	they sin,
5 ymox	caqui4ulmah ytzel	5 Imox	evil happens to them,

5 ah	rumal ytzel	5 Aj	because of evil,
	rumal ytzel ɛih		because this is an evil day,
	roo ychal.		all five.
6 queme	Ytzel ɛih	6 Kame	An evil day,
6 tihax	quehe ri	6 Tijax	like this
6 4ij	vcohom vchayl vkul	6 Tz'i'	he has placed his collar,[62]
6 yɛ	katzih	6 Iq'	truly,
6 yix	quel vi la	6 I'x	comes out there,
	hun ɛan hun rax		a yellow and a green one
	pa vchij		of his mouth;
	vtoɛom vchayl vk[u]l		he has punched his collar
	ytzel ɛih		an evil day,
	roo yichal.		and all five.
7 queh	Katzih	7 Keej	Truly,
7 cauok	v4oyim	7 Kawoq	his tangling,
7 ba4	vçisim	7 B'atz'	his trembling,[63]
7 aɛbal	vcoheic,	7 Aq'ab'al	[is] his existence,
7 4,iquin	ma na katzih	7 Tz'ikin	truly not,[64]
	ma na ka[tz]ih ta ubanom		truly not [is] his accomplishment;
	ytzel ɛih		an evil day,
	roo yichal.		all five.
8 ɛanil	Katzih	8 Q'anil	Truly,
8 hunaahpu	vtzila ɛih	8 Junajpu	a very good day.
8 ee	aree ticbal rech	8 E	This is the planting for
8 4at	alibatzil	8 K'at	daughters-in-law and
8 ahmak	a4inimakil	8 Ajmaq	noblemen[65]
	chuach ɛalibal		before the throne[66] of
	coh balam		puma and jaguar

	4o ui		they are,
	roo ychal.		all five.

/25/

9 toh	Katzih	9 Toj	Truly,
9 ymox	vtzi[l]ah ɛih	9 Imox	a very good day,
9 ah	ahaua[b]al ɛih	9 Aj	kingdom day;
9 can	ekay rech	9 Kan	the bearer for it
9 nooh	pataliy re	9 No'j	the server[67] for it,
	ah cak ah amaɛ		he of peace,[68]
	ah nima4ux		he of generosity,
	ekay rech		bearer for
	amaɛ tinamit		nation and town,
	ekay pu rech		bearer also for
	alabil a4inimakil		the young men and noblemen,
	roo ychal		all five
	vae vtzilah ɛih.		of these are very good days.
10 4,ij	Ma na vtzilah ɛih tah	10 Tz'i'	Not a very good day.
10 yɛ	xapu vij vuach 4o ui	10 Iq'	Merely above is its face,
10 yix	vnaoh	10 I'x	its meaning is
10 queme	bakil holomal	10 Kame	bone-ness and skull-ness,
10 thihax	chico[lo]tah vi	10 Tijax	it shall be saved,
	ri yavab		the sick,
	ri chi chapatah		the ones who may have been caught within this day,
	chupam vae ɛih		all five.
	roo yichal.		
11 ba4,	Katzih	11 B'atz'	Truly,

11 aɛbal	vtanabal ɛih,	11 Aq'ab'al	inquisition day.[69]
11 4,iquin	xaui 4u 4o	11 Tz'ikin	There is only
11 queh	v4axto4abal	11 Keej	deceiving[70]
11 cauok	chi ɛih	11 Kawoq	on this day,
	rumal vchapam		because he seized
	vcanab chuɛab		the prisoner by force;
	ytzel ɛih		an evil day,
	roo ychal.		all five.
12 ee	Katzih	12 E	Truly,
12 4at	yaohibal rech limosna	12 K'at	[place of] offering of alms
12 ahmak	chuach Dios nima ahauh	12 Ajmaq	before God, the great Lord;
12 ɛanil	ta hun missa	12 Q'anil	when a mass,
12 hunahpu	ta hun responso	12 Junajpu	when a responsory
	quech animas		for the souls [shall be sung], ✿
	chi 4onobex		it shall be petitioned
	rech 4azlemal		for life;
	roquibal		the entering of
	elah oɛeh		request and lament,
	oher tzih:		it is an ancient word,
	vtzilah ɛih		this is a very good day,
	roo ychal.		all five.

/26/

13 ah	Katzih	13 Aj	Truly,
13 can	royoval ɛih,	13 Kan	a day of anger,
13 noh	rumal katzih coo	13 No'j	because [it is] truly powerful,

13 thoh	xaui 4u queb catzihon vi	13 Toj	only then speak twice
13 ymox	u4ux vnaoh	13 Imox	heart and mind
	vae εih		of this day,
	roo ychal.		all five.
1 yix	Katzih	1 I'x	Truly,
1 queme	royoval εih	1 Kame	a day of anger.
1 tihax	vtiobal cumatz	1 Tijax	The bite of the serpent,
1 4,ij	vtiobal balam	1 Tz'i'	the bite of the jaguar;
1 yε	ca4hacun coh balam	1 Iq'	puma and jaguar eat flesh[71]
	xe çivan		below in the ravine,
	oher tzih.		it is an ancient word.
2 4,iquin	Katzih	2 Tz'ikin	Truly,
2 queh	vtzilah εih:	2 Keej	a very good day,
2 cauok	chirij ca4hacun vi	2 Kawoq	where [they] eat flesh
2 ba4	cot balam	2 B'atz'	eagle and jaguar
2 aεbal	chupam vae εih	2 Aq'ab'al	within this day,
	roo yi[ch]al.		all five.
3 ahmak	Ytzel εih	3 Ajmaq	An evil day,
3 εanil	ti4,il vuij	3 Q'anil	tight is his hair[72]
3 hunahpu	pa boh	3 Junajpu	in the cotton
3 ee	quehe rij	3 E	like this,
3 4at	colim kaha boh	3 K'at	loosened down is the cotton from his hair,
	pa vij		like this it says,
	quehe cubijh		all five of this day.
	roo ychal vae εih.		

/27/

4 noh	Ytzel ɛih	4 No'j	An evil day,
4 thoh	bakil holomal	4 Toj	bone-ness and skull-ness
4 ymox	vuach	4 Imox	are its aspect,
4 ah	rumal	4 Aj	because
4 can	ma na 4aceinnak ta	4 Kan	not awakened[73]
	vae ɛih		is this day,
	roo yichal.		all five.
5 tihax	Katzih	5 Tijax	Truly,
5 4,ij	royoval ɛih,	5 Tz'i'	a day of anger,
5 yɛ	chuxiquin 4oui royoval	5 Iq'	in its ear there is anger;
5 yx	xaui 4u	5 I'x	only
5 queme	cayb catzihon ui	5 Kame	twice speaks
	vae ɛih		this day,
	roo ychal.		all five.
6 cauok	Katzih	6 Kawoq	Truly,
6 ba4	chaommalah ɛih	6 B'atz'	a very beautiful day,
6 aɛbal	aree ta	6 Aq'ab'al	when
6 4,iquin	quecuchuu quib yaquiab,	6 Tz'ikin	the Yaqui assemble themselves
6 queh	chuach Dios nima ahauh	6 Keej	before God, the great Lord;
	oher tzih		it is an ancient word;
	vtzilah ɛih		a very good day,
	roo ychal.		all five.
7 hunahpu	Ytzel ɛih	7 Junajpu	An evil day,
7 ee	chuxalcatbee	7 E	on the crossroad
7 4at	vcubam vi rib	7 K'at	he seated himself,
7 ahmak	chirij cuban vi	7 Ajmaq	where he does

7 ɛanil	vpatan banom mahi cutzin ala quehe ui cabe cu' ro yichal.	7 Q'anil	his service and work, nothing does the boy well, as he leaves then, all five.

/28/

8 ymox	Katzih	8 Imox	Truly,
8 ah	utzi[l]ah ɛih,	8 Aj	a very good day.
8 can[74]	ti4il	8 Kan	It is placed
8 noh	nimac sel nima lak	8 No'j	the great bowl, the great plate
8 toh	chuach katzih chaom ɛih vtzilah roo ychal.	8 Toj	before him, truly a beautiful day, very good, all five.
9 yɛ	Katzih	9 Iq'	Truly,
9 yx	vtzilah ɛih	9 I'x	a very good day,
9 queme	vcouil ɛih	9 Kame	a powerful day,
9 tihax	chaom ɛih	9 Tijax	a beautiful day,
9 4,ij	rumal chuach ɛalibal cot balam 4o ui roo yichal va ɛih.	9 Tz'i'	because before the throne of eagle and jaguar; it is all five of this day.
10 aɛbal		10 Aq'ab'al	
10 4,iquin		10 Tz'ikin	
10 queh		10 Keej	
10 cauok		10 Kawoq	
10 ba4,		10 B'atz'	

11 4at	Katzih	11 K'at	Truly,
11 ahmak	ytzel ɛih	11 Ajmaq	an evil day,
11 canil	ma na vtzilah ɛih ta	11 Q'anil	not a very good day,
11 hunahpu	roo yichal.	11 Junajpu	all five.
11 ee		11 E	

/29/

12 can	Katzih	12 Kan	Truly,
12 noh	vtanabal ɛih	12 No'j	inquisition day.
12 toh	aree choc.	12 Toj	When shall be entered
12 ymox	4,iquin	12 Imox	the bird,
12 ah	sip yaoh	12 Aj	a[s] gift, offering
	chuach Dios nima ahauh		before God, the great lord.
	oher tzih		It is an ancient word,
	ta caban missa		when is made a Mass,
	vae puch responso		or this response,
	roo ychal vae ɛih		all five of this day.
13 queme	Ytzel ɛih	13 Kame	An evil day,
13 tihax	katzih	13 Tijax	truly,
13 4,ij	qui4ouh chup[a]m	13 Tz'i'	it passes within,
13 yɛ	huyub taɛah	13 Iq'	the mountain valley
13 yx	ta4alibetahinak	13 I'x	was tread on,
	ɛuɛi xiquin		the owl,[75]
	ahtza ahlabal		enemy and warrior,
	roo ychal vae ɛih.		all five of this day.
1 queh	Vtzilah ɛih	1 Keej	A very good day,
1 cauok	aree chinuc ui	1 Kawoq	when shall be prepared
1 ba4	4,ibanic 4otonic	1 B'atz'	writing, sculpturing,

1 aɛbal	çuanic ɛohomanic	1 Aq'ab'al	playing flute, playing the drum, singing,
1 4,iquin	bixanic	1 Tz'ikin	when also jade and money
	vue puch xit puak		should be implored
	chuach Dios nima ah[a]uh		before God, the Great Lord.
	chi4,onox vi.		
2 ɛanil	4,onobal vɛih	2 Q'anil	Petition day,
2 hunahpu	v4azlemal ticon	2 Junajpu	for the life of the planting
2 ee	chuach Dios nima ahauh	2 E	before God, the Great Lord,
2 4at	xaui yabal	2 K'at	only the presenting of
2 ahmak	sip yaoh,	2 Ajmaq	gift and offering
	chu4amovah vi (christo)		to thank Christ,[76]
	roo ychal vae ɛih.		all five of this day.

/30/

3 toh	Katzih	3 Toj	Truly,
3 ymox	vtzilah ɛih	3 Imox	a very good day,
3 ah	chaom ɛih	3 Aj	a beautiful day,
3 can	4,onobal 4azlem	3 Kan	petition for life
3 noh	chuach Dios, nima ahauh	3 No'j	before God, the great lord,
	roo yichal vae ɛih.		all five of this day.
4 4ij	Katzih	4 Tz'i'	Truly,
4 yɛ	ytzel ɛih	4 Iq'	an evil day,
4 yx	royoval ɛih	4 I'x	a day of anger,
4 queme	ma na vtzilah ɛih ta	4 Kame	not a very good day,
4 tihax	roo yichal	4 Tijax	all five.
5 ba4	Katzih	5 B'atz'	Truly,

5 aɛbal	couilah ɛih	5 Aq'ab'al	a powerful day,
5 4,iquin	royoval ɛih	5 Tz'ikin	day of anger,
5 queh	roo ychal.	5 Keej	all five.
5 cauok		5 Kawoq	
6 ee	Katzih	6 E	Truly,
6 4at	ytzel ɛih	6 K'at	an evil day,
6 ahmak	vcohom vchayl vkul	6 Ajmaq	he has placed his collar,
6 ɛanil	bakil holomal	6 Q'anil	bone-ness and skull-ness,
6 hunahpu	vuach vyabilal	6 Junajpu	its aspect of sickness,
	roo yichal vae ɛih.		all five of this day.

/31/

7 ah	Ytzel ɛih	7 Aj	An evil day,
7 can	katzih	7 Kan	truly,
7 nooh	caçoconic cak belel	7 No'j	rests[77] the serpent[78]
7 toh	xepopp	7 Toj	beneath the mat,
7 ymox	xe4hacat	7 Imox	beneath the seat,
	ma na vtzilah ɛih ta		not a very good day,
	roo ychal.		all five.
8 yx	Ma na katzih ta	8 I'x	Truly not
8 queme	vtzilah ɛih	8 Kame	a very good day,
8 tihax	xa ma na katzih ta	8 Tijax	merely truly not
8 4,ij	rihobinak v4ux	8 Tz'i'	has aged his heart
8 yɛ	chuach nabe ɛalibal	8 Iq'	before the first throne
	cot balam		of eagle and jaguar,
	roo yichal.		all five.
9 4,iquin	Katzih	9 Tz'ikin	Truly,
9 queh	royoval ɛih	9 Keej	a day of anger,

9 cauok	roo ychal.	9 Kawoq	all five.
9 ba4		9 E	
9 aɛbal		9 K'at	
10 ahmak	Vtzilah ɛih	10 Ajmaq	A very good day,
10 ɛanil	katzil	10 Q'anil	truly,
10 hunahpu	vconil ɛih	10 Junajpu	a powerful day,
10 ee	roo ychal.	10 E	all five.
10 4at		10 K'at	

/32/

11 noh	Ytzel ɛih	11 No'j	An evil day,
11 toh	hun achij hun yxok	11 Toj	a man, a woman,
11 ymox	ahmac	11 Imox	sinner,
11 ah	ytzel ɛih	11 Aj	an evil day,
11 can	roo yichal.	11 Kan	all five.
12 tihax	Katzih	12 Tijax	Truly,
12 4,ij	roquibal 4,iquin	12 Tz'i'	the offering of birds
12 yɛ	chuach nima ahauh	12 Iq'	before the great lord,
12 yix	katzih	12 I'x	truly,
12 queme	utanabal ɛih	12 Kame	inquisition day,
	roo yichal.		all five.
13 cauok	Katzih	13 Kawoq	Truly,
13 ba4	ytzel ɛih	13 B'atz'	an evil day,
13 aɛbal	ma na chacamarinak ta	13 Aq'ab'al	has not grown old
13 4,iquin	v4ux vnaoh	13 Tz'ikin	his heart, his mind,
13 queh	ytzel ɛih	13 Keej	an evil day,
	roo yichal.		all five.
1 hunahpu	Katzih	1 Junajpu	Truly,

1 ee	vtzilah εih	1 E	a very good day,
1 4at	chaom εih	1 K'at	a beautiful day,
1 ahmak	chuach εalibal	1 Ajmaq	before the throne
1 εanil	cot balam	1 Q'anil	of eagle and jaguar
	4o ui		it is,
	roo ychal.		all five.

/33/

2 ymox	Vtzilah εih	2 Imox	A very good day,
2 ah	chaom εih	2 Aj	a beautiful day,
2 can	roo yichal.	2 Kan	all five.
2 nooh		2 No'j	
2 toh		2 Toj	
3 yε	Chuvach nabe εalibal	3 Iq'	Before the first throne of
3 yix	cot balam	3 I'x	eagle and jaguar
3 queme	4o ui vi	3 Kame	it is.[79]
3 tihax	ma 4u ka[tz]ih ta	3 Tijax	Then it is truly not
3 4,ij	ubanom	3 Tz'i'	his deed
	rumal		because
	quel εan,		it comes out yellow,
	quel rax		it comes out green
	pa vchij		from his mouth,
	queeb catzihon vi		twice it speaks,
	roo ychal.		all five.
4 aεbal	Hun yxok tazul	4 Aq'ab'al	One woman—divider,[80]
4 4,iquin	hun yxok hurakan	4 Tz'ikin	one woman—one leg,[81]
4 queh	cayacou	4 Keej	she is the one who raises
4 cauok	[ç]och ε[o]hom	4 Kawoq	rattle and drum,

4 ba4	ytzel ɛih	4 B'atz'	an evil day,
	roo ychal.		all five.
5 4at	Katzih	5 K'at	Truly,
5 ahmak	vtzilah ɛih	5 Ajmaq	a very good day,
5 ɛanil	katzih	5 Q'anil	truly,
5 hunahpu	chaom ɛih	5 Junajpu	a beautiful day,
5 ee	roo ychal.	5 E	all five.

/34/

6 can	Xaui vtzilah ɛih	6 Kan	Just a very good day,
6 noh	roo ychal.	6 No'j	all five.
6 toh		6 Toj	
6 ymox		6 Imox	
6 ah		6 Aj	

7 queme	Vchapem	7 Kame	It is seized
7 tihax	vcanab vtelechee	7 Tijax	his prisoner, his captive
7 4,ij	rumal	7 Tz'i'	by him,
7 yɛ	katzih	7 Iq'	truly,
7 yx	vcouil rachahilal ɛih	7 I'x	a day of power and strength,
	roo yichal.		all five.
8 queh	Katzil	8 Keej	Truly,
8 caok	vtzilah ɛih	8 Kawoq	a very good day.
8 ba4	ti4il	8 B'atz'	It is placed
8 aɛbal	nimaçel nimalak	8 Aq'ab'al	the great bowl, the great plate,
8 4,iquin	chuach	8 Tz'ikin	before him;
	chaom ɛih		a beautiful day,

	roo ychal.		all five.
9 ɛanil	Vtzilah ɛih	9 Q'anil	A very good day,
9 hunahpu	v4aam nima ɛohom	9 Junajpu	it is taken the great drum,
9 ee	aree ta xeleçax	9 E	when was removed
9 4at	vholom	9 K'at	the head of
9 ahmak	hun nima ahauh	9 Ajmaq	one great lord,
	chui hun c[u]bal		on top of one seat,
	balam cubal		jaguar seat,
	katzih		truly,
	achahilal catzihon		strength it speaks
	roo ychal.		—all five.

/35/

10 toh	Vtzilah ɛih	10 Toj	A very good day,
10 ymox	vyacatahibal beyom	10 Imox	the raising of the trader;
10 ah	ru4aam	10 Aj	he took
10 can	vual v4hamey	10 Kan	his fan, his staff,
10 noh	chacol 4,iquin	10 No'j	perching the bird
	puij		on top of it.
	katzih		Truly,
	vtzilah ɛih		a very good day,
	roo ychal.		all five.
11 4,ij	Vcouil ɛih	11 Tz'i'	A powerful day,
11 yɛ	nabek	11 Iq'	at first
11 yix	xa aree	11 I'x	merely he
11 queme	queeb catzihon vij	11 Kame	speaks twice:
11 thihax	caban utz	11 Tijax	good is done,
	caban ytzel		evil is done,
	roo yichal.		all five.

12 ba4	Roɛibal 4,iquin	12 B'atz'	Offering of birds,
12 aɛbal	vtanabal ɛih	12 Aq'ab'al	inquisition day,
12 4iquin	vtzilah ɛih	12 Tz'ikin	a very good day,
12 queh	yabal limosna	12 Keej	the giving of alms
12 cavook	chuach Dios nima ahauh	12 Kawoq	before God, the great lord,
	roo ychal vae ɛih.		all five of this day.
13 ee	Ytzel ɛih	13 E	An evil day,
13 4ut	va4al re	13 K'at	arrayed are the teeth,
13 ahmak	va4al rix4ak	13 Ajmaq	arrayed are the claws
13 ɛanil	xo4h tucur	13 Q'anil	of owl and eagle-owl
13 hunahpu	chui	13 Junajpu	above,
	roo ychal vae ɛih.		all five of this day.

/36/

Oxlahu vinak chi ɛih.	Thirteen twenty-day [periods].
Vae ahilabal ɛih chi nima ronohel	This is the counting of all days.
varal chi4am vi naoh	Here counsel shall be taken
chirech ka4azlem kauinakirem,	as to our life and our creation
chuach Dios nima ahauh	before God, the Great Lord.
oh ral 4ual kanabe kahauh Adan	We are the children of our First Father, Adam,
oh 4ural kabe chuch eva:	we are the children of our First Mother, Eve.
oh 4o varal chuach vleuh	We live here on earth
chupam ruleval maclabal	in earthly sin.
oh 4o ui chi nima konohel	We are here in our great commune,
oh vuach vleual vinak:	we are worldly people,
oh 4u ral v4ahol	and we are the daughters and sons
ka nima ahaual Jesuchristo –	of our great lord Jesus Christ.

Are 4u vae ɛih	It is then this day,
ahilabal ɛih	the count of days,[82]
vue chi4uxlaax hun missa	when shall be remembered a Mass
chuach Dios nim ahual,	before God, the Great Lord,
vae puch missa quech animas,	or when a Mass for the souls,
vue puch sip yaoh limosna	or when presents, offerings, alms,
vue puch candela,	or when candles,
choc	shall be offered
chuach Dios nima ahauh.	before God, the great lord.
aree 4ut choc	When shall enter
ri chaomalah ɛih	the very beautiful day;
vue roobal ɛih	if the fifth day,
vue puch vuakakibal ɛih.	or if the sixth day,
Vbelehebal ɛih	the ninth day,
vcablahubal ɛih,	the twelfth day,
rutzil vchaomal ɛih,	is a day of goodness and beauty
chi4onobex	it shall be petitioned
rech 4azlemal	for life
chuach Dios nima ahauh	before God, the great lord.
chahuartah chikih	It shall be guarded our back,
oh Christianos Vinak.	of us, the Christian people.
En 6 diçiembre de 1722 años.	On December 6 of the year 1722.

Calendar B-II

/37/[83]

1 canil	Vt[z]i[l]ah ɛih	1 Q'anil	Very good day,[84]
1 hunahpu	chaom ɛih	1 Junajpu	a beautiful day,
1 eé	chu ɛalibal coh	1 E	at the puma throne,

1 4at	calibal balam	1 K'at	the jaguar throne
1 ahmak	4o in	1 Ajmaq	they are,
	ahcak ahmak		the angry one and the sinner,
	siqil [z]en		the calling, the laughing,[85]
	4o lem qo u[z].		there is order, there is good.[86]
2 toh	Vtzilah ɛih	2 Toj	Very good day,
2 ymox	xa us itzel ɛih	2 Imox	merely a good and evil day,
2 ah	caib catzihon vi	2 Aj	twice he speaks.
2 can	xa chelic	2 Kan	Merely shall come out,
2 noh	xa choc	2 No'j	merely shall enter
	vqux ahcaival		the heart of the seer,
	qux ru4		the heart with him,
	xa cuhul chi ɛoh[om]		merely the striker with the drum.[87]
3 4,ij	Xa chel	3 Tz'i'	Merely
3 yɛ	ɛan rax	3 Iq'	yellow and green
3 balam	pa uchi	3 B'alam	shall come out of his mouth,
3 came	caib chitzihon	3 Kame	twice it shall speak,
3 tihax	us ahrach	3 Tijax	the good companion.
	nabek		First there is
	ɛalibal balam		the jaguar throne,
	nabe qo		first there is
	us xaan		fly and mosquito;[88]
	xa ahchutina4ux		merely the one of small heart,[89] merely the placer of his word.

xaj ɛei utzih

4 baq	Hun yxok hun achi	4 B'atz'	A woman, a man,
4 acbal	queyacoy	4 Aq'ab'al	they raise
4 4iquin	çoch ɛohon	4 Tz'ikin	rattle and drum.
4 queh	xatuqul hun rakan	4 Keej	It rotated one leg.
4 caok	ytzel ɛih	4 Kawoq	An evil day,
	çic 4oy		the cry of the monkey
	u4oheic.		is its being.

/38/

5 e	[U]tzilah ɛih	5 E	Very good day,[90]
5 4at	chaan ɛih	5 K'at	a beautiful day,
5 ahmak	roquebal yaoh	5 Ajmaq	the entering of presents,
5 canil[91]	vtaobal ɛih	5 Q'anil	petition day.
5 hunahpu		5 Junajpu	
6 ah	[U]tzilah ɛih	6 Aj	Very good day,
6 can	roquebal yaoh	6 Kan	entering of presents,
6 noh	vtaobal ɛih	6 No'j	petition day.
6 toh		6 Toj	
6 ymos		6 Imox	
7 balam	Ki quel ral vchab vi	7 B'alam	Truly comes out the arrow;
7 came	vchapem	7 Kame	he has seized
7 tihax	vcanab vteleché	7 Tijax	his prisoner, his captive,
7 4,ij	ri vcouil	7 Tz'i'	this powerful
7 yɛ	rachahilal ɛih	7 Iq'	and strong day;[92]
	ki coc		truly enters

	v4amal vcolobal		the stringing [and] roping.[93]
8 4,iquin	vtzilah ɛih	8 Tz'ikin	A very good day.
8 quieh	tiquil	8 Keej	It is placed
8 caok	nima lak nima sel	8 Kawoq	the great plate, the great bowl
8 ba4,	chuach	8 B'atz'	before him.
8 aɛbal	xa vcochin	8 Aq'ab'al	Merely he has received
	re ahavarem		his lordship,
	4ohlem cubul		the custom of seating
	chupam ɛalibal balam		upon the jaguar throne.

/39/

9 ahmak	Ki xa	9 Ajmaq	Truly merely,
9 ɛanil	u4helem uholom 9 ɛanil	9 Q'anil	he carried in his arms his head; truly merely,
9 hunahpu	ki xa	9 Junajpu	he took his head;
9 e	ru4aam uholom 9e	9 E	about this it speaks,
9 4at	chui catzihon vi kitzih	9 K'at	truly.
10 noh	vyacatahibal	10 No'j	The raising of
10 toh	ahgai ahbeyon	10 Toj	the merchant, the trader;
10 ymos	ru4am	10 Imox	he took
10 ah	uval vghami[y]	10 Aj	his fan [and] his staff,
10 can	chacal xɛanqi 4,iquin chui rekan	10 Kan	the little yellow bird sitting upon his load.
11 tihax	vcovil ɛih	11 Tijax	A powerful day,
11 4,ij	utz itzel	11 Tz'i'	good and evil,

11 yε	caib catzihon vi	11 Iq'	twice it speaks.
11 balam		11 B'alam	
11 came		11 Kame	
12 caok	vt[a]obal εih 12 Caok	12 Kawoq	Petition day,
12 ba4,	relechibal 12 Ba4,	12 B'atz'	supplication,
12 aεbal	roquebal yaoh	12 Aq'ab'al	the entering of presents.
12 4,iquin	12 4,iquin	12 Tz'ikin	
12 queh		12 Keej	

/40/

13 hunahpu	Ytzel εih	13 Junajpu	An evil day,
13 e	baqal chic	13 E	already arrayed
13 4at	pa pop	13 K'at	on the mat,
13 ahmak	bak holom chic	13 Ajmaq	already bone and skull
13 εanil	uvach	13 Q'anil	is its aspect,
	xachácal		merely shall pass over,
	[x]ic tucur pa uvi		hawk[94] and owl above,
	ximil che		fastened [are]
	uεab raεan		his arms and legs.
1 ymos	vtzilah εih	1 Imox	Very good day,
1 ah	ticbal ave[x]abal	1 Aj	for planting and sowing.
1 can		1 Kan	
1 noh		1 No'j	
1 toh		1 Toj	
2 yε	Ytzel εih	2 Iq'	An evil day,
2 balam	uεih eleεom	2 B'alam	the day of the thief.
2 came	xa xere	2 Kame	Merely only,

2 tihax	xa chiquis pa uvi	2 Tijax	merely shall it end above,
2 4,ij	xa chilisah chui	2 Tz'i'	merely shall it release above,
	relee mesonel		the theft of the sweeper
	vqoheic		[is] his existence.
	xa		Merely
	chi ya chi siuan		in the river,[95] in the ravine
	chiqohe		shall it be,
	chi ui		where
	vqux xa chelic		his heart shall merely come out.[96]
3 aebal	vtzilah eih	3 Aq'ab'al	Very good day,
3 4,iquin	nabek	3 Tz'ikin	at first,
3 queh	xa xere	3 Keej	merely only,
3 caok	xauí xatirinak uqoheic	3 Kawoq	only was adjusted his existence, merely [this] shall be returned.
3 ba4,	xa chitzelexic	3 B'atz'	

/41/

4 4at	Ki roio[v]al eih	4 K'at	Truly, a day of anger;
4 amak	ki xa ratixam vquiquel	4 Ajmaq	truly, merely he sneezed blood;
4 eanil	mana chicovinic	4 Q'anil	he shall not get strong.
4 hunahpu	qui nabe ro[i]eval eih	4 Junajpu	Truly, first day of anger,
4 e	xare xa chiquis pa uvi	4 E	only merely shall it end above.
5 cam	Ki roieval eih	5 Kan	Truly, a day of anger,

5 noh	vtz itzel	5 No'j	good and bad,
5 toh	vqhacubal	5 Toj	place where eagle and
5 ymos	cot balam	5 Imox	jaguar eat flesh,
5 ah	xe tak çivam	5 Aj	below in the ravines,
	ruq ki ahmak		and, truly, the sinner,
	hun yxok hun achi		a woman, a man,
	4ehoxovic		they fornicate.
6 came	Ma na chícovinic	6 Kame	It shall not get strong,
6 tihax	ru4 xa chel	6 Tijax	and merely shall come out
6 4,ij	ɛan rax	6 Tz'i'	yellow and green
6 yɛ	pa vchí	6 Iq'	from his mouth,
6 balan	ruq xa ah ixokichinel	6 B'alam	and merely the blasphemous,[97]
	xa chitzuvnic		merely he shall watch.
	xa pu chicaic		And merely he shall observe.
7 queh	Ki nabe vtzilah ɛih	7 Keej	Truly, first very good day,
7 caok	vcouil ɛih puch	7 Kawoq	also a powerful day,
7 baq	ki gu	7 B'atz'	truly, then
7 aɛbal	xa vgoim rib	7 Aq'ab'al	merely he has tangled himself, merely he has made his planting.
7 4,iquin	xa vbanom vtiquic	7 Tz'ikin	

/42/

8 ɛanil	Vtzilah ɛih	8 Q'anil	A very good day,
8 hunahpu	are chivan vi	8 Junajpu	this is when shall be sown daughters-in-law and noblemen, before the throne of

8 e	alibatzil aqinimakil	8 E	puma and jaguar.
8 4at	chuach ɛalibal	8 K'at	
8 ahmak	coh balam	8 Ajmaq	
9 toh	vtzilah ɛih	9 Toj	A very good day,
9 ymos	ek[ay] huiub	9 Imox	bearer[98] of mountains
9 ah	ekaleí amaɛ	9 Aj	bearer of nation,
9 can	tinamit sivan	9 Kan	of town and ravine;
9 noh	vuaibal ruqabal ɛih	9 No'j	day for eating and drinking,
	roquebal ah4hab		the entering of the shooter,
	relebal ahqhab		the exit of the shooter.
10 4,ij	Ytzel ɛih	10 Tz'i'	An evil day,
10 yɛ	bakil holomal	10 Iq'	bone-ness and skull-ness
10 balam	chigohe vi	10 B'alam	shall be there;
10 came	xavi nabe qut	10 Kame	only first therefore
10 tihax	chicolotah	10 Tijax	it shall be saved,
	ri iab		the sick
	chichapahtah		shall not be caught
	chi yavabil		in sickness
	chupam		within [these five days].
11 ba4,	ki vqaxtoqabal ɛih	11 B'atz'	Truly, day of deceiving,
11 aɛbal	ki vchapem	11 Aq'ab'al	truly, he seized [him]
11 4,iquin	chuɛul vɛab	11 Tz'ikin	by the wrist of his hand,
11 queh	ki chi hul	11 Keej	truly into the abyss,
11 caok	ki chi çiauan	11 Kawoq	truly into the ravine,
	taqal vi kahok		straight downward.

/43/

12 e	Chaom εih	12 E	Beautiful day,
12 4at	are [c]hi qhacun vi	12 K'at	when eat flesh
12 ahmak	cot balam	12 Ajmaq	eagle and jaguar.
12 εanil	vtaobal εih	12 Q'anil	Petition day,
12 hunahpu	may puch, xoc vi	12 Junajpu	and also entered
	tol		Tol,[99]
	ral		Ral,[100]
	tapi cholol		Tapi Cholol,[101]
	xcox		the parrot,[102]
	εan q,iquin		the yellow bird;
	loεbal ça[k]amaε		love and peace,
	ruq releh chí		and supplication,
	roquibal yaoh		the entering of presents.
13 ah	Roieval εih	13 Aj	A day of anger,
13 can	maui 4o ta retabal	13 Kan	not is there a sign,
13 noh	ma pu go ta	13 No'j	there also is no[thing]
13 toh	cucolo	13 Toj	that he defends,
13 ymos	xa humul vtz humul itzel	13 Imox	merely once good, once evil.
1 balam	Ki royeval εí	1 B'alam	Truly, a day of anger,
1 came	vcouil εih	1 Kame	a powerful day,
1 tihax	vtiobal	1 Tijax	the bite of
1 4,ij	cumatz balam	1 Tz'i'	serpent and jaguar,
1 yε	maui chicolotah chutio	1 Iq'	not shall it be saved, shall it bite.
2 4,iquin	Ytzel εih	2 Tz'ikin	An evil day,

2 queh	ma qu ɛecabinak	2 Keej	not yet has it become old,[103]
2 caok	xa tiquil	2 Kawoq	merely planted
2 ba4,	vui chi çivan	2 B'atz'	at the top in the ravine,
2 aɛbal	xa pu	2 Aq'ab'al	and merely
	vɛoɛam vkul		he has struck his throat
	chi qhab		with an arrow;
	ytzel ɛih		an evil day,
	royeval ɛih.		a day of anger.

/44/

3 Ahmak	Chaom ɛih	3 Ajmaq	A beautiful day,
3 ɛanil	vɛih abix	3 Q'anil	day of the milpa,
3 hunahpu	vɛih ticon	3 Junajpu	day of the planting,
3 e	ticbal avexabal	3 E	planting and sowing,
3 4at	ruq xavi	3 K'at	with only
	u4hacubal		the place where eagle
	cot balam		and jaguar eat flesh,
	vtzilah ɛih		a very good day.
4 noh	Ytzel ɛih	4 No'j	An evil day,
4 toh	bakil holomal	4 Toj	bone-ness and skull-ness
4 ymos	chiqolic	4 Imox	will be there,
4 ah	ma na ɛeçabinak	4 Aj	not has it become old
4 can	ki	4 Kan	truly,
	nabe navinak		the first diviner.
5 tihax	royeval ɛih	5 Tijax	A day of anger,
5 4,ij	xa ahcakuach	5 Tz'i'	merely he of hate,

5 yɛ	ahmosvach	5 Iq'	he of jealousy,
5 balam	xa xiquin vuach	5 B'alam	merely listening,[104]
5 came	chimukun vi	5 Kame	he shall see,
	caib vach chitzihon vi		two aspects he shall speak,
	humul tael		once the one who hears,
	humul maui tael		once the one who does not hear.
6 caok	Chaom ɛih	6 Kawoq	A beautiful day,
6 ba4,	are quequlum vi quib	6 B'atz'	when assemble themselves,
6 aɛbal	yaquiab	6 Aq'ab'al	the Yaqui people,
6 4,iquin	are chi ban vi 4hab	6 Tz'ikin	when shall be made the arrow, when shall enter
6 queh	are choc vi	6 Keej	great prosperity[105]
	nima[l]ah loɛobal		[and] peace.
	çak amak		

7 hunahpu	Ytzel ɛih	7 Junajpu	An evil day,
7 e	cahib vxalcatbe	7 E	four crossroads
7 4at	chubeyah	7 K'at	he shall walk,
7 ahmak	xa chícuxcu[n] chirí	7 Ajmaq	merely he shall chew there
7 ɛanil	xa paxinak uqux	7 Q'anil	merely broken is his heart.
8 ymos	Vtzilah ɛih	8 Imox	Very good day,
8 ah	chaomalah ɛih	8 Aj	very beautiful day,
8 can	are chahilax vi	8 Kan	when shall be counted

8 noh	chol ɛih	8 No'j	the order of days,
8 toh	maí ɛih	8 Toj	the prophecy[106] of days,
	cumal ahauab		by the lords for
	ticbal auexabal		planting and sowing
9 yɛ	Vcouil ɛih	9 Iq'	Powerful day,
9 balam	vɛalibal coh	9 B'alam	puma throne,
9 came	ɛalibal balam	9 Kame	jaguar throne,
9 tihax	ɛalibal cot	9 Tijax	eagle throne,
9 4,ii	qo ui	9 Tz'i'	there is;
	xaxe ki		merely truly,
	royeval ɛih		a day of anger,
	ri chichapatah		who shall be caught
	chupan		within
	chi iavabil		in sickness.
10 aɛbal	R[u]qam	10 Aq'ab'al	He took
10 4,iquin	usach uɛohom	10 Tz'ikin	his rattle, his drum;
10 queh	x4,ul	10 Keej	the Xtz'ul dance,
10 caok	loqom paytatom	10 Kawoq	the well liked, the buffoon[107]
10 ba4,	v4oheic	10 B'atz'	is his existence.
	xa chuya		Merely he shall give
	vçoch vɛohon		his rattle, his drum,
	xa tzeletzih		merely the joker.[108]

/46/

11 4at	Ytzel ɛih	11 K'at	Evil day,
11 ahmak	tiquil	11 Ajmaq	it is planted
11 ɛanil	chi hul chi çivan	11 Q'anil	in the abyss, in the ravine,

11 hunahpu	xa pu benak v4ux	11 Junajpu	also merely went his heart
11 e	chi biz chi m[e]εen	11 E	in sorrow, in heat,
	xa vqison vqux		merely has finished his heart,[109]
	xa utz biz.		merely good [and] sorrow.[110]
12 can	[Ut]aobal εih	12 Kan	Petition day,
12 noh	chi chi elehebal εih	12 No'j	in the end of day,[111]
12 toh	4,apibal	12 Toj	the closure,
12 ymos	vui hon	12 Imox	the top of the yard,[112]
12 ah	hul çivan	12 Aj	abyss, ravine;
	xavi roquebal yaoh		only the entering of presents.
13 came	Ytzel εih	13 Kame	An evil day,
13 tihax	xa chiqou	13 Tijax	merely shall it pass
13 4,ii	chupam vbe	13 Tz'i'	within the path of
13 yε	εuεi tuεur 4habi tuεur	13 Iq'	quetzal-owl, arrow-owl,
13 balam	4hala4hoh vlok	13 B'alam	listening[113] hither
	vxiguin tucur		the owl's ear is
	chupam vui vbee		on the head of his path.
1 queh	Vtzilah εih	1 Keej	A very good day.
1 caok	are chinuc vi	1 Kawoq	It shall be prepared
1 ba4,	bixanam	1 B'atz'	the singing,
1 aεbal	are chinuc vi	1 Aq'ab'al	it shall be prepared
1 4,iquin	çu vεohon	1 Tz'ikin	flute and drum,
	4ot 4,ib		sculpturing and writing,
	puakinic		silver-making,[114]

quemenic baq,inic			weaving and spinning;
utzilah ɛih			a very good day,
na vinak ɛih			not a human day,
vɛovil ɛih			a powerful day.

/47/

2 ɛanil	4ulbal	2 Q'anil	The reunion,
2 hunahpu	uɛih abix ticon	2 Junajpu	day of milpa and planting,
2 e	vuebal ru4obal ɛih	2 E	day of eating and drinking,
2 4at	ɛanalah ɛuih	2 K'at	very yellow day,
2 ahmak	raxalah ɛih	2 Ajmaq	very green day.
3 toh	Vtzilaj ɛih	3 Toj	Very good day,
3 ymos	ticbal auexabal	3 Imox	for planting and sowing,
3 ah	çak amaɛ cauexic	3 Aj	peace is sown.
3 can	chaɛan nima 4humil	3 Kan	It shall ascend the great star,
3 noh	eko ɛih	3 No'j	the morning star,[115]
	chaon ɛih		a beautiful day.
4 4,ii	Royeval ɛih	4 Tz'i	Day of anger,
4 yɛ	caib catzihon vi	4 Iq'	twice it speaks,
4 balam	vtiobal	4 B'alam	the bite of
4 came	cumatz balam	4 Kame	serpent and jaguar.
4 tihax		4 Tijax	
5 ba4,	royeval ɛih	5 B'atz'	Day of anger,
5 aɛbal	ki che ki 4am	5 Aq'ab'al	truly wood, truly cord
5 4,iquin	v4ux	5 Tz'ikin	is his heart;
5 queh	vɛovil ɛih chic	5 Keej	a powerful day already,

5 caok	abah vuach	5 Kawoq	stone is its aspect,
	utz ytzel		good and evil.

/48/

6 e	Xa uɛoɛam ukul	6 E	Merely he has struck his throat
6 4at	chi 4hab	6 K'at	with an arrow,
6 amak	xa	6 Ajmaq	merely
6 ɛanil	toɛol rib	6 Q'anil	the one who hurts himself,
6 hunahpu	xa pu	6 Junajpu	or merely
	camisal rib		the one who kills himself.
	qo nabe		There is first
	vua ruquiaax		his food [and] his drink,
	are xa qhutin uqux		it is merely small his heart,
	ruq xavi qate		and only then
	chelech chin chic		he shall surrender[116] again,
	humul chic		once again,
	rah chic v4açeic		his wish is to live again.[117]
7 ah	Ytzel ɛih	7 Aj	An evil day,
7 can	ki xacho qou	7 Kan	truly merely shall it pass,
7 noh	çucun cakabelom	7 No'j	seated the serpent[118]
7 toh	xe vtem	7 Toj	beneath the bench,
7 ymos	xa vpop	7 Imox	beneath the mat,
	xe v4hacat		beneath the seat,
	chupam ube		within its path;

	roqebal		entering of
	muk çiçon		woodworm and weevil;[119]
	are xmukur		it is the dove
	vui caka che		on the top of the red tree.
8 balam	Caib catzihon vi	8 B'alam	Twice it speaks,
8 came	qhakap urihil	8 Kame	partly his elders,
8 tihax	4hacap ral 4alal	8 Tijax	partly his offspring,
8 4,ii	ma na ki rihobinak	8 Tz'i'	truly not aged,
8 yε	xa 4hutin v4ux	8 Iq'	merely small is his heart.
	ki nabe		Truly, first
	couilah εih		powerful day,
	4hacap rax		partly green,[120]
	4hakap chaεih		partly dry.
9 4,iquin	Royeual εih	9 Tz'ikin	Day of anger,
9 queh	nabek ene	9 Keej	at first [—],[121]
9 caok	xaui rutzil	9 Kawoq	only his kindness,
9 ba4,	xa vcouil rachahilal	9 B'atz'	merely his power and strength;
9 aεbal	roquebal	9 Aq'ab'al	the entering of
	ah 4hab ah pocob		the archer, the shield-bearer;
	vtaobal εih		petition day.

/49/

10 ahmak	Vcovuil εih	10 Ajmaq	A powerful day,
10 canil	vcouil vuach	10 Q'anil	powerful is its aspect,
10 hunahpu	vεih chuchuxinak	10 Junajpu	day of being taken mother,[122]

10 e	kahavinakak	10 E	fathering[123]
10 4at	v4oheic	10 K'at	is its existence.
11 noh	Hun yxok, hun achi 11 noh	11 No'j	A woman, a man,
11 toh	quealaxic	11 Toj	they are born,
11 ymos	cakinac	11 Imox	reddened,
11 ah	maui xaɛilic	11 Aj	not is it detained,
11 can	maui xacataxic chicah 11 ymos[124] 11 ah 11 can	11 Kan	not is it impeded [from] above.
12 tihax	Vcouil ɛih 12 tihax	12 Tijax	A powerful day,
12 4,ij	xa xe	12 Tz'i'	merely below,
12 yɛ	e meçobal 12 yɛ	12 Iq'	they are brooms
12 balam	vui hom 12 balam	12 B'alam	[on] the top of the yard,
12 came	vtaobal ɛih roquebal yaoh	12 Kame	petition day, the entering of offerings.
13 caok	Ytzel ɛih	13 Kawoq	An evil day,
13 ba4,	chee 4aam v4ux	13 B'atz'	wood and cord are his heart
13 aɛbal	chi nima ahauarem	13 Aq'ab'al	in the great reign;
13 4,iquin	maui 4o ta	13 Tz'ikin	there is no
13 queh	vtobal ucolobal chuach e abah catzihon vi mavi chutao.	13 Keej	help, salvation before those who are stone, he speaks, he shall not hear.

Calendar C

/50/

Vahxaklahuqal ruq hoob ɛih ri hun hunab.	Eighteen [times] twenty with five days of one year.
1. Chee 20	Che' 20
2. Tequexpual 20	Tekexepoal 20
3. 4,ibapop 20	Tz'ib'a Pop 20
4. Çac 20	Saq 20
5. 4hab 20	Ch'ab' 20
6. Mam 20	Mam 20
7. Vcab mam 20 5 ɛih chiroyobeh choc ahaval.	Ukab' Mam 20; five days when waiting for the entering of the regent.
8. Liquin ca 20	Likinka 20
9. Vcab liquin ca 20	Ukab' Likinka 20
10. Pach 20	Pach 20
11. Vcab Pach 20	Ukab' Pach 20
12. 4,içilakan 20	Tz'isi Laqam 20
13. 4,iquin ɛih 20	Tz'ikin Q'ij 20
14. Cakam 20	Kaqam 20
15. Botam 20	B'otam 20
16. Çih 20	Si'j 20
17. Vcab çih 20	Ukab' Si'j 20
18. Vrox çih 20	Urox Si'j 20
4,apiɛih 400	Closing Days

FOUR

Calendario de Vicente Hernández Spina, 1854

A third calendar was recorded by the resident priest of Santa Catarina Ixtlahuacán, Vicente Hernández Spina. Pedro Cortés y Larraz (1958, 2:150) describes the eighteenth-century village of Santa Catarina Ixtlahuacán:

> The town of Santa Catarina, which they speak about in the parish of San Miguel Totonicapan, is located in a ravine at the base of high mountains. The houses are covered with terra-cotta tiles, there are no streets, and the entire area is surrounded by ravines. They grow many potatoes, wheat, and maize, and have commerce with the coast, the Indians are very rich, but with bad reputations of being very vicious. They burn copal most of the time, have zarabandas, drunkedness, and changing women, contribute that in the same house live three or four families together. It is very common. In a short period of time they have built a church, with many statues of saints, and ornamentos of silver . . . none of them are buried in the church because the doors of the church are closed. At the edge of the town there is a small sweat bath, where they are reduced to promiscuity because men, women, boys and girls enter together at all hours.

In 1854, the German physician Karl Scherzer visited Santa Catarina Ixtahuacan, hoping the community had preserved much of its traditional culture because of its isolated location. The resident priest, Fray Vicente Hernández Spina, prepared an informative document of K'iche' religion and language in anticipation of Scherzer's visit. It included a general account of the religious concepts of the Ixtahuacanos, their calendar day names, an explanation of how the calendar worked, a recorded prayer offered by a K'iche' shaman (written in K'iche' and Spanish), a list of important sacred places and the names of shamans in the area, and a small K'iche' grammar. Scherzer published the vocabulary in 1855 and also published the K'iche' prayer along with his own commentaries on religion at Ixtahuacan (1856, 1864, 1954).

The French abbé Charles Etienne Brasseur de Bourbourg (1857, 1:lxxxvi) obtained this manuscript from the chief archivist in Guatemala City shortly thereafter, and near the end of the nineteenth century it became part of the Brinton Collection. A partial copy of it was later translated into English by Ethel Bunting and published in the *Maya Society Quarterly* (1932).

The document begins with a title page (f. 1) and an introduction (ff. 2–3), followed by three calendar wheels (ff. 4–6; Figs. 4.2–4.4). The Hernández calendar lists the twenty day names with their associated good and bad fates (f. 8). The month names are not given, but Hernández indicates that there were eighteen twenty-day units. He also gives the beginning point of the K'iche' solar year in relation to the Christian calendar (on May 1) and indicates the year-bearer pattern (No'j, Iq', Kej, E) (f. 7; Fig. 4.5). In addition, on folios 9–10, Hernández recorded in K'iche' a long invocation given by a shaman (*ajq'ij*) of Santa Catarina and translated it into Spanish. He also included a brief summary of indigenous beliefs and customs, a list of sacred toponyms, and the names of shamans in the region (f. 12a) and their shrines (f. 12b).

Hernández must have interrogated several important shamans of the community in order to learn about their calendar. One must have trusted him because he permitted Hernández to record the prayer in its entirety and then revealed the list of important shamans residing in the area.

In recent years the community has experienced political and social turmoil because of its relocation after a mudslide. A government-sponsored geological report in the 1950s warned of the instability of the Santa Catarina Ixtlahuacán area and the risk that the town could slide down the mountain to the creek below. This happened in November 1998, when the population of Santa Catarina suffered the devastation of Hurricane Mitch and recurrent earth tremors. Two people died, and much of the town was later demolished by swollen rivers and mudslides.

/1/ Kalendario conçervado hasta el dia por los Sacerdotes del Sol en Ixtlavacam, pueblo descendiente de la Nación Kiché, descubierto por el Presbítero Vicente Hernández Spina, San Catarina Yxtlavacam, Agosto 19 de 1854.

/2/ La poblacion de Ixtlavacam compuesta de ventiun mil almas, dista doce leguas S.O. de la antigua capital del Quiché, en Centro America, y tres al S. de la cabezera del Departamento Totonicapam.

Ningun vestígio, ningun monumento recuerda la existencia de una poblacion anterior á la que hoy existe.

Muerto Tecum Umam á manos del Gral. Español Dn. Pedro de Alvarado; subyugada la Nacion, la Capital del reyno Quiché debía sufrir los vel vejámenes anejos á la suerte que cabe á los vencidos. Gran parte de la poblacion, queriendo evitar los males que tenia emigró del Quiché, y buscaba en los montes contíguos un lugar en donde vivir independiente, y segura de la persecucion.

Los Yxtlauacanes tomando el desierto que hoy se llama el alto de Totonicapam, encontraron una hoya profunda é inaccessible, cuyo fondo cortado con profundas barrancas y lleno de riscos, fué elegide para su habitacion, sirviendoles las cuevas que se encontraban bajo los peñascos enormes, en lugar de casas.

El zelo de los primeros misioneros apostólicos descubrió esta Colonia; y tentando las dificultades para trasladarlos á major lugar, convinieron en edificar en una maseta, que se encuestra al medio de esta profundidad, un Templo bajo de la advocacion del Santa Catarina Martir.

La situacion topográfica del pueblo, colocaba á los Yxtlauacanes en su verdadero aislamiuento. Solo robustecidos por la aspereza del lugar, dedicados esclusivamente á la agricultura, fieles á sus matrimónios que contrahen casi en la infancia, se multiplicaron, y permanecieron en un estado verdaderamiente independiente; y han pasado con su genio, su religion y sus costumbres primitivas, al través de los siglos y en medio de sus conquistadores.

Conocer bien los Yxtlahuacanes, es haber conocido la Nacion Quiché. Si la religion es un sentimiento natural, la Nacion Quiché no carecía de el al tiempo de él al tiempo de la Conquista.

La existencia de los Sacerdótes, que en su idioma, y en el numero singular se llaman Aj-quij; y la tabla con sus correspondientes signos, demuestran esta verdad. Dividen el Gobierno del mundo entre dos princípio sigualmente poderosos. Uno perfectamente bueno, que habíta en las Alturas; y otro malo, dueño de la tierra.

Calendar Preserved to the Present Day by the Priests of the Sun[1] in Ixtlahuacán, a Town that Descends from the K'iche' Nation, Discovered by the Priest, Vicente Hernández Spina, Santa Catarina Ixtlahuacán, August 19, 1854.[2]

The population of Ixtlahuacán is 21,000 souls. The town lies twelve leagues southwest of the ancient K'iche' capital in Central America, and three leagues south of Totonicapan.

No vestige or monuments record the existence of any inhabitants prior to those of today.

After Tecum Uman had been killed by the Spanish conqueror Pedro de Alvarado, and the K'iche' nation subdued, the capital had to suffer the usual fate and its consequent oppressions. To evade these, a large part of the population emigrated to the south to the mountains' fastnesses where they might live independent and free from persecution.

The Ixtlahuacáns took themselves to what is known as the Totonicapan highlands, where they found a deep and almost inaccessible ravine cut into steep barrancas and filled with crags. This they chose for their home and in the caves under the great rocks they made their dwellings.

The place was later discovered by the zeal of the early missionaries; but faced with the difficulty of trying to have the Ixtlahuacáns removed to some more accessible location, they arranged to have built on a small prominence in the middle of the ravine a church dedicated to Santa Catarina Martir.

The topography of the place gave true isolation to the inhabitants. Living to themselves, made robust by the very austerities of their surroundings, dedicated exclusively to agriculture, loyal to marriages that were contracted almost in their infancy, they multiplied in a state of real independence. They brought down with them through the centuries their own spirit, their religion and its primitive customs in the very midst of their conquerors.

To know the Ixtlahuacáns is to have known the K'iche' nation; and if we think of religion as a natural sentiment, the K'iche's of the conquest were not lacking therein.

The existence of their priests, whom they called *ajq'ij*, and their calendrical tables with their particular signs confirm this statement. They regarded the world as under the domination of two equally powerful principles: one wholly good, living in the heights, and the other evil and master of the earth.

	1	2	3	4	5	6	7	8	9	10	11	12	13	14	15	16	17	18
Kawoq	1	2	3	4	5	6	7	8	9	10	11	12	13	14	15	16	17	18
[Jun]ajpu		1	2	3	4	5	6	7	8	9	10	11	12	13	14	15	16	17
Imox			1	2	3	4	5	6	7	8	9	10	11	12	13	14	15	16
Iq'				1	2	3	4	5	6	7	8	9	10	11	12	13	14	15
Aq'ab'al					1	2	3	4	5	6	7	8	9	10	11	12	13	14
K'at						1	2	3	4	5	6	7	8	9	10	11	12	13
Kan							1	2	3	4	5	6	7	8	9	10	11	12
Keme								1	2	3	4	5	6	7	8	9	10	11
Keej									1	2	3	4	5	6	7	8	9	10
Q'anil										1	2	3	4	5	6	7	8	9
Toj											1	2	3	4	5	6	7	8
Tz'i'												1	2	3	4	5	6	7
B'atz'													1	2	3	4	5	6
E														1	2	3	4	5
Aj															1	2	3	4
I'x																1	2	3
Tz'ikin																	1	2
Ajmaq																		1
N'oj																		
Tijax																		

4.1. SEQUENCE OF DAY NAMES AND NUMBERS, F. 3, 1722 K'ICHE' CALENDAR.

4.2. DRAWING OF CALENDAR WHEEL, F. 4, 1854 K'ICHE' CALENDAR.

```
20
19  20
18  19  20
17  18  19  20
16  17  18  19  20
15  16  17  18  19  20
14  15  16  17  18  19  20
13  14  15  16  17  18  19  20
12  13  14  15  16  17  18  19  20
11  12  13  14  15  16  17  18  19  20
10  11  12  13  14  15  16  17  18  19  20
9   10  11  12  13  14  15  16  17  18  19  20
8   9   10  11  12  13  14  15  16  17  18  19  20
7   8   9   10  11  12  13  14  15  16  17  18  19  20
6   7   8   9   10  11  12  13  14  15  16  17  18  19  20
5   6   7   8   9   10  11  12  13  14  15  16  17  18  19  20
4   5   6   7   8   9   10  11  12  13  14  15  16  17  18  19  20
3   4   5   6   7   8   9   10  11  12  13  14  15  16  17  18  19  20
2   3   4   5   6   7   8   9   10  11  12  13  14  15  16  17  18  19  20
```

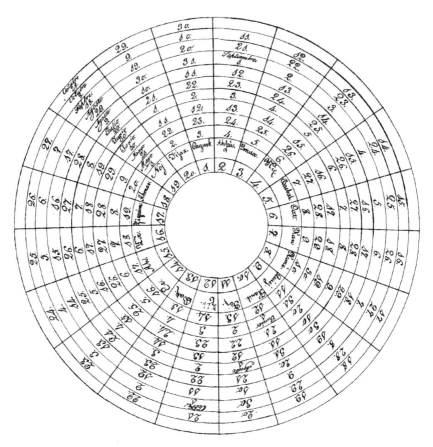

4.3. DRAWING OF CALENDAR WHEEL, F. 5, 1854 K'ICHE' CALENDAR.

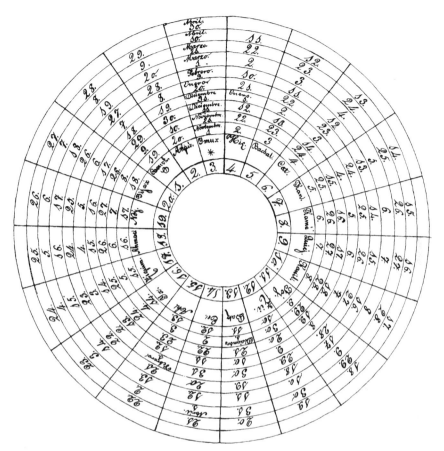

4.4. DRAWING OF CALENDAR WHEEL, F. 6, 1854 K'ICHE' CALENDAR.

Mayo	Junio	Julio	Agosto	Setiembre	Octubre
1 No'j	1 Q'anil	1 Tijax	1 Toj	1 Junajpu	1 Tz'i'
2 Tijax	2 Toj	2 Kawoq	2 Tz'i'	2 Imox	2 B'atz'
3 Kawoq	3 Tz'i'	3 Junajpu	3 B'atz'	3 Iq'	3 E
4 Junajpu	4 B'atz'	4 Imox	4 E	4 Aq'ab'al	4 Aj
5 Imox	5 E	5 Iq'	5 Aj	5 K'at	5 I'x
6 Iq'	6 Aj	6 Aq'ab'al	6 I'x	6 Kan	6 Tz'ikin
7 Aq'ab'al	7 I'x	7 K'at	7 Tz'ikin	7 Kame	7 Ajmaq
8 K'at	8 Tz'ikin	8 Kan	8 Ajmaq	8 Keej	8 No'j
9 Kan	9 Ajmaq	9 Kame	9 No'j	9 Q'anil	9 Tijax
10 Kame	10 No'j	10 Keej	10 Tijax	10 Toj	10 Kawoq
11 Keej	11 Tijax	11 Q'anil	11 Kawoq	11 Tz'i'	11 Junajpu
12 Q'anil	12 Kawoq	12 Toj	12 Junajpu	12 B'atz'	12 Imox
13 Toj	13 Junajpu	13 Tz'i'	13 Imox	13 E	13 Iq'
14 Tz'i'	14 Imox	14 B'atz'	14 Iq'	14 Aj	14 Aq'ab'al
15 B'atz'	15 Iq'	15 E	15 Aq'ab'al	15 I'x	15 K'at
16 E	16 Aq'ab'al	16 Aj	16 K'at	16 Tz'ikin	16 Kan
17 Aj	17 K'at	17 I'x	17 Kan	17 Ajmaq	17 Kame
18 I'x	18 Kan	18 Tz'ikin	18 Kame	18 No'j	18 Keej
19 Tz'ikin	19 Kame	19 Ajmaq	19 Keej	19 Tijax	19 Q'anil
20 Ajmaq	20 Keej	20 No'j	20 Q'anil	20 Kawoq	20 Toj
21 No'j	21 Q'anil	21 Tijax	21 Toj	21 Junajpu	21 Tz'i'
22 Tijax	22 Toj	22 Kawoq	22 Tz'i'	22 Imox	22 B'atz'
23 Kawoq	23 Tz'i'	23 Junajpu	23 B'atz'	23 Iq'	23 E
24 Junajpu	24 B'atz'	24 Imox	24 E	24 Aq'ab'al	24 Aj
25 Imox	25 E	25 Iq'	25 Aj	25 K'at	25 I'x
26 Iq'	26 Aj	26 Aq'ab'al	26 I'x	26 Kan	26 Tz'ikin
27 Aq'ab'al	27 I'x	27 K'at	27 Tz'ikin	27 Kame	27 Ajmaq
28 K'at	28 Tz'ikin	28 Kan	28 Ajmaq	28 Keej	28 No'j
29 Kan	29 Ajmaq	29 Kame	29 No'j	29 Q'anil	29 Tijax
30 Kame	30 No'j	30 Keej	30 Tijax	30 Toj	30 Kawoq
31 Keej		31 Q'anil	31 Kawoq		31 Junajpu

Noviembre		Deciembre		Enero		Febrero		Marzo		Abril	
1	Imox	1	B'atz'	1	Iq'	1	Aj	1	Imox	1	E
2	Iq'	2	E	2	Aq'ab'al	2	I'x	2	Iq'	2	Aj
3	Aq'ab'al	3	Aj	3	K'at	3	Tz'ikin	3	Aq'ab'al	3	I'x
4	K'at	4	I'x	4	Kan	4	Ajmaq	4	K'at	4	Tz'ikin
5	Kan	5	Tz'ikin	5	Kame	5	No'j	5	Kan	5	Ajmaq
6	Kame	6	Ajmaq	6	Keej	6	Tijax	6	Kame	6	No'j
7	Keej	7	No'j	7	Q'anil	7	Kawoq	7	Keej	7	Tijax
8	Q'anil	8	Tijax	8	Toj	8	Junajpu	8	Q'anil	8	Kawoq
9	Toj	9	Kawoq	9	Tz'i'	9	Imox	9	Toj	9	Junajpu
10	Tz'i'	10	Junajpu	10	B'atz'	10	Iq'	10	Tz'i'	10	Imox
11	B'atz'	11	Imox	11	E	11	Aq'ab'al	11	B'atz'	11	Iq'
12	E	12	Iq'	12	Aj	12	K'at	12	E	12	Aq'ab'al
13	Aj	13	Aq'ab'al	13	I'x	13	Kan	13	Aj	13	K'at
14	I'x	14	K'at	14	Tz'ikin	14	Kame	14	I'x	14	Kan
15	Tz'ikin	15	Kan	15	Ajmaq	15	Keej	15	Tz'ikin	15	Kame
16	Ajmaq	16	Kame	16	No'j	16	Q'anil	16	Ajmaq	16	Keej
17	No'j	17	Keej	17	Tijax	17	Toj	17	No'j	17	Q'anil
18	Tijax	18	Q'anil	18	Kawoq	18	Tz'i'	18	Tijax	18	Toj
19	Kawoq	19	Toj	19	Junajpu	19	B'atz'	19	Kawoq	19	Tz'i'
20	Junajpu	20	Tz'i'	20	Imox	20	E	20	Junajpu	20	B'atz'
21	Imox	21	B'atz'	21	Iq'	21	Aj	21	Imox	21	E
22	Iq'	22	E	22	Aq'ab'al	22	I'x	22	Iq'	22	Aj
23	Aq'ab'al	23	Aj	23	K'at	23	Tz'ikin	23	Aq'ab'al	23	I'x
24	K'at	24	I'x	24	Kan	24	Ajmaq	24	K'at	24	Tz'ikin
25	Kan	25	Tz'ikin	25	Kame	25	No'j	25	Kan	25	Ajmaq
26	Kame	26	Ajmaq	26	Keej	26	Tijax	26	Kame	26	No'j
27	Keej	27	No'j	27	Q'anil	27	Kawoq	27	Keej	27	Tijax
28	Q'anil	28	Tijax	28	Toj	28	Junajpu	28	Q'anil	28	Kawoq
29	Toj	29	Kawoq	29	Tz'i'			29	Toj	29	Junajpu
30	Tz'i'	30	Junajpu	30	B'atz'			30	Tz'i'	30	Imox
		31	Imox	31	E			31	B'atz'		

	January		February		March		April		May		June
1	I'x	1	Kan	1	Aj	1	K'at	1	I'x	1	Kan
2	Tz'ikin	2	Kame	2	I'x	2	Kan	2	Tz'ikin	2	Kame
3	Ajmaq	3	Kej	3	Tz'ikin	3	Kame	3	Ajmaq	3	Kej
4	No'j	4	Q'anil	4	Ajmaq	4	Kej	4	No'j	4	Q'anil
5	Tijax	5	Toj	5	No'j	5	Q'anil	5	Tijax	5	Toj
6	Kawoq	6	Tz'i"	6	Tijax	6	Toj	6	Kawoq	6	Tz'i"
7	Ajpu'	7	B'atz'	7	Kawoq	7	Tz'i"	7	Ajpu'	7	B'atz'
8	Imox	8	E	8	Ajpu'	8	B'atz'	8	Imox	8	E
9	Iq'	9	Aj	9	Imox	9	E	9	Iq'	9	Aj
10	Aq'ab'al	10	I'x	10	Iq'	10	Aj	10	Aq'ab'al	10	I'x
11	K'at	11	Tz'ikin	11	Aq'ab'al	11	I'x	11	K'at	11	Tz'ikin
12	Kan	12	Ajmaq	12	K'at	12	Tz'ikin	12	Kan	12	Ajmaq
13	Kame	13	No'j	13	Kan	13	Ajmaq	13	Kame	13	No'j
14	Kej	14	Tijax	14	Kame	14	No'j	14	Kej	14	Tijax
15	Q'anil	15	Kawoq	15	Kej	15	Tijax	15	Q'anil	15	Kawoq
16	Toj	16	Ajpu'	16	Q'anil	16	Kawoq	16	Toj	16	Ajpu'
17	Tz'i"	17	Imox	17	Toj	17	Ajpu'	17	Tz'i"	17	Imox
18	B'atz'	18	Iq'	18	Tz'i"	18	Imox	18	B'atz'	18	Iq'
19	E	19	Aq'ab'al	19	B'atz'	19	Iq'	19	E	19	Aq'ab'al
20	Aj	20	K'at	20	E	20	Aq'ab'al	20	Aj	20	K'at
21	I'x	21	Kan	21	Aj	21	K'at	21	I'x	21	Kan
22	Tz'ikin	22	Kame	22	I'x	22	Kan	22	Tz'ikin	22	Kame
23	Ajmaq	23	Kej	23	Tz'ikin	23	Kame	23	Ajmaq	23	Kej
24	No'j	24	Q'anil	24	Ajmaq	24	Kej	24	No'j	24	Q'anil
25	Tijax	25	Toj	25	No'j	25	Q'anil	25	Tijax	25	Toj
26	Kawoq	26	Tz'i"	26	Tijax	26	Toj	26	Kawoq	26	Tz'i"
27	Ajpu'	27	B'atz'	27	Kawoq	27	Tz'i"	27	Ajpu'	27	B'atz'
28	Imox	28	E	28	Ajpu'	28	B'atz'	28	Imox	28	E
29	Iq'			29	Imox	29	E	29	Iq'	29	Aj
30	Aq'ab'al			30	Iq'	30	Aj	30	Aq'ab'al	30	I'x
31	K'at			31	Aq'ab'al	31		31	K'at		

July	August	September	October	November	December
1 Tz'ikin	1 Kame	1 No'j	1 Kej	1 11 Tijax	1 2 Q'anil
2 Ajmaq	2 Kej	2 Tijax	2 Q'anil	2 12 Kawoq	2 3 Toj
3 No'j	3 Q'anil	3 Kawoq	3 Toj	3 13 Ajpu'	3 4 Tz'i"
4 Tijax	4 Toj	4 Ajpu'	4 Tz'i"	4 1 Imox	4 5 B'atz'
5 Kawoq	5 Tz'i"	5 Imox	5 B'atz'	5 2 Iq'	5 6 E
6 Ajpu'	6 B'atz'	6 Iq'	6 E	6 3 Aq'ab'al	6 7 Aj
7 Imox	7 E	7 Aq'ab'al	7 Aj	7 4 K'at	7 8 I'x
8 Iq'	8 Aj	8 K'at	8 I'x	8 5 Kan	8 9 Tz'ikin
9 Aq'ab'al	9 I'x	9 Kan	9 Tz'ikin	9 6 Kame	9 10 Ajmaq
10 K'at	10 Tz'ikin	10 Kame	10 Ajmaq	10 7 Kej	10 11 No'j
11 Kan	11 Ajmaq	11 Kej	11 No'j	11 8 Q'anil	11 12 Tijax
12 Kame	12 No'j	12 Q'anil	12 Tijax	12 9 Toj	12 13 Kawoq
13 Kej	13 Tijax	13 Toj	13 Kawoq	13 10 Tz'i"	13 1 Ajpu'
14 Q'anil	14 Kawoq	14 Tz'i"	14 Ajpu'	14 11 B'atz'	14 2 Imox
15 Toj	15 Ajpu'	15 B'atz'	15 Imox	15 12 E	15 3 Iq'
16 Tz'i'	16 Imox	16 E	16 Iq'	16 13 Aj	16 4 Aq'ab'al
17 B'atz'	17 Iq'	17 Aj	17 Aq'ab'al	17 1 I'x	17 5 K'at
18 E	18 Aq'ab'al	18 I'x	18 K'at	18 2 Tz'ikin	18 6 Kan
19 Aj	19 K'at	19 Tz'ikin	19 Kan	19 3 Ajmaq	19 7 Kame
20 I'x	20 Kan	20 Ajmaq	20 Kame	20 4 No'j	20 8 Kej
21 Tz'ikin	21 Kame	21 No'j	21 Kej	21 5 Tijax	21 9 Q'anil
22 Ajmaq	22 Kej	22 Tijax	22 Q'anil	22 6 Kawoq	22 10 Toj
23 No'j	23 Q'anil	23 Kawoq	23 Toj	23 7 Ajpu'	23 11 Tz'i"
24 Tijax	24 Toj	24 Ajpu'	24 Tz'i"	24 8 Imox	24 12 B'atz'
25 Kawoq	25 Tz'i"	25 Imox	25 B'atz'	25 9 Iq'	25 13 E
26 Ajpu'	26 B'atz'	26 Iq'	26 E	26 10 Aq'ab'al	26 1 Aj
27 Imox	27 E	27 Aq'ab'al	27 Aj	27 11 K'at	27 2 I'x
28 Iq'	28 Aj	28 K'at	28 I'x	28 12 Kan	28 3 Tz'ikin
29 Aq'ab'al	29 I'x	29 Kan	29 Tz'ikin	29 13 Kame	29 4 Ajmaq
30 K'at	30 Tz'ikin	30 Kame	30 9 Ajmaq	30 1 Kej	30 5 No'j
31 Kan	31 Ajmaq		31 10 No'j		31 6 Tijax

Creen en la immortalidad del alma; pero de una manera enteramente material. Reconócen otros genio subalternos á estoy dos princípios que gobiernan el mundo, asociandose á ellos las almas de lo Ajquij y de las personas respetables de sus antepesado.

/3/ Muestra 1 Kalendario Gentilico Kichi.

/4/ Muestra 2

/5/ Muestra 3

/6/ Muestra 4

/7/ Muestra 5

/8/ Division de los Dias en Buenos, Malos é Indiferentes.

Noj, Dia bueno, consagrado al genio que preside al alma. En él se pide el buen entendimiento para el suplicante y para su família.

Tijax, Dia bueno, lo mismo que el anterior.

Caguok, Dia indiferente.

Ajpú, Indiferente.

Ymux, Dia malo. Los Sacerdotes del Sol (Ajquijes) piden á los génios malos la infelicidad para los enemigos del suplicante.

Yε, Dia malo. Es igual al anterior.

Bacbal, Malo. Ocurren los Ajquijes á sus adoratorios á pedir el mal para su enemigos.

Cat, Dia malo; igual á los anteriores.

Kan, Yd.

Kamé, Malo; igual á los antecedents.

Quiej, Dia bueno. Se pide todo lo que se provechoso al suplicante.

Kanil, Dia bueno, consagrado á los genios que presiden la agricultura. Se pide en él todo lo que sirve de sustento para el hombre.

They believe in the immortality of the soul, although in a manner wholly material. They acknowledge other lower genera below the two great ruling principles, and with these genera they also associate the souls of their *ajq'ij* and of their honored ancestors.

Example 1. K'iche' Calendar [Fig. 4.1]

Example 2. [Calendar wheel] [Fig. 4.2]

Example 3. [Calendar wheel] [Fig. 4.3]

Example 4. [Calendar wheel] [Fig. 4.4]

Example 5. [K'iche' Calendar] [Fig. 4.5]

Division of the Days into Good, Bad and Indifferent

No'j, a propitious day, dedicated to the presiding genius of the soul. On this day they pray that the suppliant and his family may be endowed with good judgment.

Tijax, good, the same as the preceding.

Kawoq, indifferent day.

Junajpu, indifferent day.

Imox, bad day; the priests of the sun, the ajq'ijab, on this day pray to the spirits of evil against their enemies.

Iq', bad day, the same as preceding.

Aq'ab'al, bad day; the ajq'ijes seek the shrines against their enemies.

K'at, bad day, the same as the preceding.

Kan, bad day, the same as the preceding.

Kame, bad day, the same as the preceding.

Keej, good day, on which beneficial things are asked for the suppliant.

Q'anil, good day, sacred to the spirits of agriculture; on this [day] are supplicated all those things that serve man's sustenance.

Toj, Dia malo. Infelíz el hombre que nace en el' Por un destino inevitable debe forzosamente ser perverso.

Tzii, Dia malo. Se pide la infelicidad para los enemígos del suplicante.

Batz, Dia malo. Los Sacerdotes piden la enfermedad; pero especialmente la paralysis para los enemígos del suplicante.

Ee, Dia bueno. En él se comienzan los contratos matrimoniales; precediendo muchas oblaciones á los genios benignos.

Aj, Dia bueno, consagrado á los genios de la agricultura; y á los que presiden á los rebaños y animals domésticos.

Yx, Dia bueno, consagrado á los genios que dominant en los montes. En él se suplica á los mismos genios contengan á los lobos y demas bestias carnivores para que no destruyan sus rebañmos y animals domésticos.

Tziquin, Dia exelentisimo. En él se hacen dobles ofrendas; en la Iglesia al Dios bueno y Supremo, y á los Santos que hay en el Templo; y en las cuevas, barrancas profundos, bosques espesos y sombríos; y en la cima de todos los montes. En él se pide todo lo que es provechoso és interesante al hombre; el perdón de sus faltas cometidas contra los dos principios Bueno y Malo; se concluyen los contratos matrimoniales; y se dá princípio á toda obra importante.

Ajmac, Dia tamien exelentisimo, lo mismo que el anterior; y consagrado ademas, á los genios que presiden á la salud.

/9/ Deprecación de un Aj Quij ó Sacerdote del Sol en el adoratorio de Raxquin en favor de N. Tzep.

Alal nu Dios Jesucristo, xa jun cuyá ri Dios cajauixel, ri Dios cajolaxel, ri Dios Espiritu Santo chupam güe quij hora, güe Tijax, saj ca lac locolaj tac animas aj relebal quij, chu cajibal quij ruc santos animas, saj cu lá lal rajagual-güinaquil, sija raxquim, saj culá santos animas Juan Vachiac, D.n Domingo Vachgiac, Juan Yxquiaptap, santos animas Fran.co Ecoquij santos animas, Diego Soom santos animas, Juan Tay santos animas, Alonso Tzep santos animas, Diego Tziquin santos animas, Don Pedro Noj alac ajquijap, alac aj punto, al rajagual, at güinaquil, at Dios juyup, Dios tacaj Don Purupeto Martin casic la ri pom candela casic la camíc, saj cu lá nu locolaj chuch Santa María, ruc lá nu cajau Señor de las Esquipulas, Señor de la Capetagua, lal nu locolaj Maria Chianta, María Xan Lurenso, María Dolores, María Santa Ana, María Tiburcia, María Carma, ajau San Miguel Algancia, Capitan Santiago, San Cristoval, San Sebastian, Simigulás, San Aventura, San Pernatin, San Andres, Santa Domas, San Burtolomé, lal nu loco

Toj, bad day; unfortunate he who is born thereon; by inevitable destiny he is doomed to be perverse.

Tz'i', bad day; on it is sought the undoing of one's enemies.

B'atz', bad day, on which sicknesses, and particularly paralysis, is prayed to fall on one's enemies.

E, good day; on this day contracts of marriage are entered into, preceded by many sacrifices to the benign powers.

Aj, again a good day, and also consecrated to the gods of agriculture and to those presiding over the flocks and domestic animals.

I'x, good day; this day is sacred to the spirits of the mountains and forests; on it protection is sought for their flocks and animals at the favor of those spirits who rule over the wolves and other carnivorous beasts.

Tz'ikin, most excellent of days; on this day double offerings are made in the church of the good and supreme deity and to the saints in the churches; also offerings [are made] in the caves, the profound barrancas, and in deep and somber woods. On this day they pray for all that is beneficial and useful to man; also pardon for all sins against the two great powers, the Good and the Evil. This is also the day for the conclusion of marriage contracts and for the beginning of all important affairs.

Ajmaq, also a most excellent day, the same as the one before. It is also especially consecrated to the spirits presiding over good health.

Invocation by an Ajq'ij or Priest of the Sun at the Shrine of Rax K'im in Favor of N. Tzep.[3]

O, my Jesus Christ, only one knows,[4] the God Father, the God Son, the Holy Spirit. On this day, at this hour, on this day Tijax, come then, You, the Holy Souls Lord Sun-Rising, Mother Sun-Setting with the Holy Souls; come then, Thou, Thou Lord of Mankind, Siha Raxquim; come then, Thou, Holy Souls of Juan Vachiac, of Don Domingo Vachiac, of Juan Yxquiaptap, the Holy Souls of Francisco Ecoquij,[5] the Holy Souls of Diego Soom, the Holy Souls of Juan Tay, the Holy Souls of Alonzo Tzep, the Holy Souls of Diego Tziquin,[6] the Holy Souls of Don Pedro Noh,[7] You, Masters of the Day, You, Masters of the Point, You, Lord You, Mankind, You, Lord of the Mountains, Lord of the Plains, Don Purupeto Martin, Thou, smell this incense, candle. Thou, smell it today. Come then, Thou, My Holy Mother Holy Mary, with Thou, My Lord of Esquipulas, Lord of Capetagua, Thou, My Holy María de Chiantla, María San Lorenzo, María Dolores, María Santa Ana, María Tiburcia, María del Carmen, Lord of San Miguel Arcangel, Capitan Santiago, San Cris

/10/ Traduccion

Oh! Jesucristo mi Dios: tú Dios hijo, con el Padre y el Espiritu Santo, eres un solo Dios. Hoy, en esta dia, a esta hora en este dia de Tijax; imploro á las santas que acompañan la aurora y los últimos rayos del dia; con las Santas almas te invóco á tí! Oh! Principe de los génios que habítas en este monte de Sija-Raxquim. Venid almas santas de Juan Vachiac, de Dn. Domingo Vachiac, de Juan Yxquiaptap; almas santas de Grancisco Ecoquij, de Diego Soom, de Juan Tay, de Alonso Tzep; almas santas repíto, de Diego Tziquin y de D.n Pedro Noj; vosotros ! Oh Sacerdotes, vostros á quienes está todo patente; y tú Principe de los génios; vosotros Dios del monte, Dios del llano D.n Purupeto Martin, venid, recibíd este incienso, recibid ahora esta candela.

Veníd tambien madre mía Sta. María y tambien mi Señor de las Esquipulas, el Sr. de Capetaguna, la muy amada María de Chiantla, con la que existe en San Lorenzo, tambien María de los Dolores, María Santa Ana, María Tiburcia, María del Carmen, con el Sr. San Miguel Arcangel, el Capitan Santiago, San Cristoval, San Sebastian, San Nicolas, San Buenaventura, San Bernardino, San Andres, Santo Tomas, San Bartolomé, y tu mi amada madre Sta. Catarina, tú amada María de Concepción, María del Rosario; Tú Señor y Rey Pascual está aquí presentes: y tú yelo, tu viento exelente, tú Dios llano, tú Dios Quiac-basulup, tu Señor de Retal-uleu, tú Sr. de San Gregório, tu dueño de Chij-masá, lal ajau Sto. Tomas, tú Sr. de San Antonio, tu Señor de Chua-tulul, tú Sr. de Samayac; lal ajau Chui-poj, lal ajau Parraché, tu Señor de Chui-pecul, lal ajau Xexac, lal ajau Chua-jolom, lal ajau Pachip, lal ajau Pasaqui-juyup, lal ajau Chui-caxtum, tu dueño de Chui-ixcanul, tu dueño de Chui-lajuj-juyup, tu dueño de Chui-ixcabint, tu dueño de

/11/ Nombre de los montes, bosques, llanuras y barrancas que se nombran en la deprecasion anterior.

Sija-Raxquim, Es la altura mas culminante que hay en los terrenos de Sta. Catarina, hacia al Norte de dicho pueblo.

Quiacbasulup, Es el cerro que divide los mojones de Santa Catarina y Santa Clara, hacia al Sur. Tambien le dicen Ajau por antonomasia.

Retaluleu, Un bosque que divide estoy terrenos de los de Santo Tomas, hacia al sud-oeste.

San Gregório, Otro bosque que se ha formado en donde existió el pueblo de este nombre. E contiguo á San Miguelito.

Chua-tulul, Montaña inmediata á Santo Tomas Suchitepequez.

Translation[94]

O, Jesus Christ, my God, Thou God the Son, with the Father and the Holy Spirit, are the only God. Today, on this day, at this hour, on this day Tihax, I call upon the Holy Souls that accompany the sun-rising and the sun-setting of the day; with these Holy Souls I call upon Thee, O Prince of the Spirits, Thou who dwells in this mountain of Siha Raxquin; come, Ye Holy Spirits of Juan Vachiac, of Don Domingo Vachiac, of Juan Ixquiaptap, the Holy Souls of Francisco Excoquieh, of Diego Soom, of Juan Tay, of Alonzo Tzep; I call the Holy Souls of Diego Tziquin and of Don Pedro Noh, You, O Priests, to whom all things are revealed, and Thou, Chief of the Spirits, You, Lords of the Mountains, Lords of the Plains, Thou, Don Purupeto Martin, come, accept this incense, accept today this candle.

Come also, My Mother Holy Mary, My Lord of Esquipulas, the Lord of Capetagua, of the Beloved María de Chiantla, with her who dwells at San Lorenzo, and also María de los Dolores, María Santa Ana, María Tiburcia, María del Carmen, with San Miguel Arcangel, Capitan Santiago, San Christoval, San Sebastian, San Nicolas, San Buenaventura, San Bernardino, San Andres, Santo Tomas, San Bartolome, and Thou My Beloved Mother Santa Catarina, Thou Beloved María de Concepción, María del Rosario, Thou Lord and King Pascual, be here present. And Thou, Frost, and Thou, Excellent Wind, Thou, God of the Plain, Thou, God of Quiac-Basulup, Thou, God of Retal-Uleu, Thou, Lord of San Gregorio, Thou, Lord of Chii-Masa, Thou, Lord of Santo Tomas, Thou, Lord of San Antonio, Thou, Lord of Chua-tulul, Thou, Lord of Samayac; Thou, Lord of Chui-poj, Thou, Lord of Parraché, Thou, Lord of Chui-pecul, Thou, Lord of Xexac, Thou, Lord of Chua-jolom, Thou, Lord of Pachip, Thou, Lord of Pasaqui-juyup, Thou, Lord of Chui-caxtum, Thou, Master of

Names of the Mountains, Forests, Valleys, and Ravines That Are Named in the Preceding Invocation.

Sija Rax K'im,[95] this is the highest height in the lands of Sta. Catarina, north of the town.

Kyaq Pa Tzulub', this is the mountain that divides the boundaries of Santa Catarina and Santa Clara, southward. It is also called Ajaw for antonomasia.

Retal Ulew, a forest that divides the lands of those of Santo Tomas, to the southwest.

San Gregório, another forest that has grown where there once was a town of that name. It is adjacent to San Miguelito.

Chuwa Tulul, a mountain range close to Santo Tomas Suchitepequez.

Chij-masa, Un rio que divide estos terrenos de los de Zunil en el punto de Santo Tomas.

Parraché, Una montaña que esta al pié del volcan de Zunil en la Costa de Suchitepequez.

Chui-pecul, El mismo volcan de Zunil por al lado que enfrente á la Costa.

Xexac, Unas lomas montañosa inmediatas al Parraché.

Chua jolom, Un lugar que está en la medianía del volcan del Zunil, en donde existe una cabeza de figura humana de piedra.

Chui ixcabiut, El cerro de Santa Maria.

Chui-sac-quichij, Un volcan immediate á Soconusco, visible desde la altura de Raxquim.

Chui-Nimajuyup, El volcan de Tacana.

Xe-tzalamchoj, La llanura de Urbina.

Chui-ucup-cruz, Un bosque que esta en las siete cruces al oeste de este pueblo.

Chui-lajuj-juyup, El volcan de Quesaltenango.

Los de mas nombres en donde suenan San Antonio, Samayac, y demas pueblos, debe enten derse que son sus bosques inmediatos. Los que no se espresan y están escrítos en la deprecación, son los bosques, barrancos y montañas del mismo terreno de Sta. Catarina; entendiendose que con dichos nombres están per sonificados los genios ó dioses tutelares de dicho lugares.

/12/ Lista de los Aj Quijes ó Sacerdotes del Sol,

con espresion de los lugáres en que viven.

En San Miguelito. Miguel Toom, ex-Gobernador; Francisco Yxquiaptap, ex-alcalde; Domingo Chox, Miguel Yxquiaptap, Juan Sac, Francisco Giatzy, Jose Xtos.

Santa Catarina. Alonso Tum, Miguel Sac, Francisco Och y Juan Ixquiaptap.

Chui quisic. Manuel Vicente y Diego Con.

Chirij-peacul. Juàn Chox ó Tziquin, Francisco Carrillo y José Carrillo.

Pacajá. Francisco Coty y Lorenso Coty.

Racan-tacaj. Pascual Tay, Juan Tay y Ramon Tzep.

Quiaca-siguan. Juan Tuney y Francisco Soom.

Pa-chipac. Diego Xtos, Diego Tambrij, y José Xtos.

Chi Masa', a river that divides the lands of those of Zunil on the edge of Santo Tomas.

Pa Ra(x) Che', a mountain at the foot of the volcano of Zunil in the coast of Suchitepequez.

Chuwi' Pekul, the same volcano of Zunil from the side that is facing the coast.

Xe' Xak, a few mountainous slopes close to Parraché.

Chuwa Jolom, a place in the midst of the volcano of Zunil where there is the head of a stone statue.

Chuwi' Ix Kab' Yut, the mountain of Santa Maria.

Chuwi' Saq K'ichi', a volcano close to Soconusco, visible from the height of Raxquim.

Chuwi' Nima Juyub', the volcano of Tacana.

Xe' Tzalam Choj, the plain of Urbina.

Chuwi' Ukub' Krus, a forest that is at the Seven Crosses west of this town.

Chuwi' Lajuj Juyub', the volcano of Quezaltenango.

The other names that make up San Antonio, Samayac, and the other towns can be understood as close forests. Those that are not explained and are written in the invocation are forests, ravines, and mountains in the lands of Sta. Catarina; understanding that these names are personifying the spirits or tutelary gods of the mentioned places.

List of the Aj Q'ijes or Priests of the Sun,

with a declaration of the places where they live.

In San Miguelito. Miguel Tum (ex-governor), Francisco Yxquiaptap (ex-mayor), Domingo Chox, Miguel Yxquiaptap, Juan Sac, Francisco Giatzy, Jose Xtos.

Santa Catarina. Alonso Tum, Miguel Sac, Francisco Och, and Juan Ixquiaptap.

Chuwi' K'isik.[96] Manuel Vicente and Diego Con.

Chiri' Pe Akul.[97] Juan Chox or Tziquin, Francisco Carrillo and José Carrillo.

Pa Kaja.[98] Francisco Coti and Lorenso Coti.

Raqan Taq'aj.[99] Pascual Tay, Juan Tay, and Ramon Tzep.

Kyaqa Siwan. Juan Tuney and Francisco Soom.

Pa Ch'ipaq.[100] Diego Xtos, Diego Tambrij, and José Xtos.

Xe patuj. Cruz Sac, Cristival Marroquin, y Francisco López.

Simajutin. Manuel Perechú.

Piximbal. Pascual Xocol.

Sempoal. Ali-Chian, Sacerdotisa.

Chui-cuil. Baltasar Ixquiaptap y Na-Paquisis Sacerdostisa.

Chij-rexóm. Francisco Tzoc.

Xe chojojché. Manuel Tziquin y Juan Varchaj.

Nombres de los Adoratorios

Los adoratorios son en todos los cerros, barrancas, bosques sombríos, cuevas, picos peligrosos y tgeneralmente todo lugar sublime. Pero los adoratorios conocidos son: Sija Raxquim; Siete Cruces; Tzibaché; Sempoal.

Y un lugar en los límites con la hacienda de Argueta. En San Miguelito, bajo todo arbol grandoso; siendo preferable la Seiba.

Los instrumentos de adivinacion, son los prismas; y todo crystal que pueda partir la luz en los siete colores primoridales. A estos se agregan unos granos de tzité ó píto, mesclados los negros con los tintos para significar major los dias Buenos y malos.

/13/ Notas

1. Se ha dicho antes que los Ixtlauacanes adoran dos principios; bueno y malo. El bueno está reprersentado en el astro que preside el dia, padre de la luz y fecundador de la tierra; en su idioma se llaman Quij. El malo es el Juyúp dueño de todas las riquezas del mundo; su representación es la misma que la figura humana, la mas horrible que ellos pueden imaginar; omnipotente para favorecer con los bienes de la tierra á sus adoradores; y para dañar á los que reusan su adoración.

2. In muestra 1a contiene el sistema religiosa de los Ixtlauacanes. La posicion particular en que está, indica la sucesión continua de sus meses de veinte dias. Se llama Calendario por que cada dia es el primero del mes que indíca el signo.

Xe' Pa Tuj.[101] Cruz Sac, Cristival Marroquin, and Francisco López.

Tzima Jutin. Manuel Perechú.

Pa Ximb'al.[102] Pascual Xocol.

Sempoal. Ali-Chian, priestess.

Chuwi' Kuwil.[103] Baltasar Ixquiaptap and Na-Paquisis, priestess.

Chi Rexom.[104] Francisco Tzoc.

Xe' Chojoj Che'.[105] Manuel Tziquin and Juan Varchaj.

Names of the Shrines

The shrines are in all of the mountains, ravines, shaded forests, caves, dangerous summits, and generally in all lofty places. The known shrines are Sija Rax K'im, Siete Cruces,[106] Tzib'a Che',[107] Sempoal.

And a place within the limits of the Argueta hacienda. In San Miguelito, under a tall tree that is most likely a ceiba.

The implements of divination are prisms and all crystals capable of separating light into the seven primordial colors. To these they add a few grains of tzite or pito mixed with the black [beans] and the colored [beans] to indicate the good and bad days.

Examples

1. As I have said above, the Ixtlahuacáns worship two great principles, of the Good and of the Evil. The Good is represented by the Sun who presides over the day, the Father of Light, and Fructifier of the Earth, which in their language is called *q'ij*. The Evil Spirit is *juyub'*, master of the riches of the world; he is represented in human form, but horrible beyond imagination; he is omnipotent to favor his worshippers with earthly goods and to harm those who refuse him worship.

Since then they adhere to and recognize these two equally independent and sovereign powers, they follow the same division of good and evil gods or subordinate spirits; and for an equal reason the days assigned to all these are counted as either good or bad in their omens.

2. The tables and plans I have given contain their religious system, as above. The days in succession form a calendar made up of months of twenty days each, the days as shown in the tables marking the beginning of these months. The year in this religious system has no relation to the astronomical year but serves only to indicate the destiny or lot that falls to each man, bound absolutely to the day whereon he is born. This count they carry forward with scrupulous accuracy.

3. Como en este sistema religioso gentílico quiche, su año ninguna relación tiene con el año astronómico; sino que únicam.te sirve para señalar la suerte que le toca á cada hombre, ligada absolutam.te. Al dia en que nace, llevan una cuenta escrupulosa y exacta de los dias y meses de su signo. Para esto sirve la muestra 2a esplicada de esta manera.

Hoy dia 12 de agosto, en nuestro Calendario, es el 1o de Ajpu para el Sacerdote del Sol; 20 del Caguoc, 19 de Tihax, 18 de Noj, 17 de Ajmac, 16 de Tziquin, 15 de Yx, 14 del mes Aj, 13 de Ee, 12 de Batz, 11 de Tzii, 10 de Toj, 9 de Kanil, 8 de Kiej, 7 de Kamé, 6 de Kan, 5 de Cat, 4 de Bacbal, 3 de Yε, y 2 de Ymux; contándose hácia la izquierda. Los Yxtlavacanes no han tenido ningun signo ó carácter para pintar sus ideas; es decir, no tienen escritura. Asi és, que los Sacerdotes que llevan sin equiocarse la cuenta de los dias Buenos y malos, de sus meses ó signos, atribuyendoles sus respectivas influencias buenas ó malas, gozan entre sus conciudadanos de tanta consideración.

4. Ya se ha dicho que el año Quiché ninguna relación tiene con el año astronómico. Se dice impropiam.te año por una fiesta que celebrant annualmente al princípio de mayo, ó á mediaio de la primavera.

En este año de 1854 comensó el año religioso cabalistico de los Ixtlauacanes con el signo Noj que corresponde al 1o de mayo, como se vé en la muestra 3a que comprende los meses del mismo mayo, junio, Julio, agosto, setiembre y octubre en este primer hemisfério; y en la muestra 4a los meses de noviembre, diciembre, enero, febrero, marzo y abril.

5. Los Sacerdotes (Ajquijap) tienen dos maneras de contra su año; Primera, dandole á cada signo un solo dia, y se vé en la plana 5a, que el signo Noj, Tijax, Caguoc, Ajpú, Ee é Ymux entran en el año diez y nueve veces cado uno; y que el signo Yε hasta el Ajmac, entran en el año gentílico diez y ocho veces, componiendo los viente signos del almanaque desde 1o de mayo hasta el 30 de abril, vispera de su gran fiesta gentílica, la suma de trescientos sesenta y cinco dias.

La segunda manera de contra es la de darle á cada signo un mes de veinte dias; y asi van sucesivamente alternandose hasta completer su año de dies y nueve meses, y lo mismo Tijax, Caguoc, Ajpú, Ee é Ymux; y de diez y ocho meses los signos desde Yε hasta el Ajmac.

Sirve tambien la plana 5a para que el Parroco indágue a punto fijo el dia en que ocurren los tales Sacerdotes á la Yglesia; que son los dias exelentísimos de Ajmac, Tziquin y Kanil.

3. In this religious system of the K'iche' type, the year is not related to the astronomical year but serves exclusively to indicate the fate that affects every human being, who is absolutely tied to it. On the day he is born, they carry out a scrupulous and exact calculation of the days and months that are his sign. This is the purpose of the [Fig. 4.2] which can be explained like this:

Today, the 12th of August in our calendar [in 1854], is the first day of Junajpu for the priest of the Sun; after it follow in succession, 20 Kawoq, 19 Tijax, 18 No'j, 17 Ajmaq, 16 Tz'ikin, 15 I'x, 14 Aj, 13 Ee, 12 Batz, 11 Tzii, 10 Toj, 9 Kanil, 8 Keej, 7 Kamé, 6 Kan, 5 K'at, 4 Aq'ab'al, 3 Iq', and 2 Imox, counting to the left. The Ixtlahuacáns have no signs or characters wherewith to paint their ideas, that is, no writing. Thus it is that the priests who carry on without error the counting of these good and evil days of their months, assigning to each its respective good and evil influences, enjoy the fullest consideration among their own people.

4. I spoke above of the astronomical year, which is inexact in this respect, in that their "year" begins on the first of May, in the middle of spring.

In this year 1854 the cabalistic year begins, on May 1, with the sign No'j, as can be seen in the [Fig. 4.3], which comprises the months of the same year May, June, July, August, September, and October, in this first hemisphere and in the example the months of November, December, January, February, March, and April.

5. The priests, the ajq'ijab', have two ways of counting their year. First, matching each sign with one single day, as can be seen in [Fig. 4.5] [we see] that the signs No'j, Tijax, Kawoq, Junajpu, E, and Imox occur nineteen times within one year; and that the signs Iq' up to Ajmaq occur eighteen times in the local year, combining the twenty signs of the almanac starting on the 10th of May up to the 30th of April, which is the eve of the main local festival; the sum is 365 days.

The second way of counting (time) is to match each sign with a month of twenty days; like this follow nineteen months sucessively alternating until the year is completed; and the same Tijax, Kawoq, Junajpu, E, and Imox and of eighteen months the sign from Iq' up to Ajmaq.

[Fig. 4.5] indicates to the priest the point the day when the same ajq'ijab' come to the church, being on the most excellent days, Ajmaq, Tz'ikin, and Q'anel.

Tan puntualmente observan los dioas de su Calendario, que mientras los dias Domingos y fiestas solemnes de nuestra religion se halla el Templo desierto, en los dias Buenos de su almanáque, se halla llena de adoradores que queman insienso, llenan todo el pavimento de caldelas y cantando responses; con cuyas apariencias católicas han engañado por luego tiempo á los Parrocos que ignoraban su idioma, y que no podian observer la mescal impía de los nombres cristianos con los de sus Númenes y Manes de sus antepasados sacerdotes, como se vé en la Deprecación y traducción literal.

So punctually do they observe these days of their calendar, that whereas the Sundays and solemn fiestas of our religion find the church empty, on the good days of their almanac they are crowded with worshippers who burn incense, filling the pavement with candles and chanting the responses. With these Catholic appearances they have long deceived the parish curates who did not understand their language and could not note the impious mingling of Christian names with those of the spirits and names of their ancient priests, as appears in the adjoined invocation and literal translation.

APPENDIX ONE

Notes on Highland Maya Calendars
ROBERT BURKITT, ca. 1920

The calendar is everywhere the one calendar, but with endless variations; and you meet those variations not only as you might, perhaps, expect in passing from language to language but constantly, in passing from town to town. One town of a language preserves more of the calendar, and another less; or preserves different elements of it, or has different superstitions about it; or, finally, one town of a language preserves the calendar, while another knows nothing about it, and the language is in places where the calendar is not. The Indian calendar has been preserved or lost in the same way as other Indian customs, and the languages themselves have been preserved or lost, not language by language, but town by town; and the calendar which is preserved in seven or eight languages is in fact the independent and various possession of some thirty or forty towns. (Burkitt 1930–1931:107)

Robert Burkitt (1869–1945) was one of the few ethnographers working in highland Guatemala during the first decades of the twentieth century. He was familiar with the Q'eqchi' language and gathered important texts and documents written in that language during his lengthy residence in the northern Maya highlands of the Department of Alta Verapaz. Burkitt's work, although linguistically oriented, contains considerable ethnographic and archaeological information. He published various extensive transcriptions of Q'eqchi' texts, as well as lists of surnames, a sixteenth-century will, folktales, and a calendar from Solomá in northwestern Guatemala.

The notes presented here were discovered by Dr. Elin Danien in December 1979 in a wooden crate among Burkitt's possessions in a storage building at Finca Sepacuite, east of Senahu in the Department of Alta Verapaz (Danien 1985). There is no title page or explanation for the text although "Dres. Ms." (Dresden Manuscript) is used as a header on each page of writing. The paper measures five by eight inches, and the pagination is continuous from 3567 to 3605, indicating that these notes originally must have been included within a much larger manuscript. These were Burkitt's personal notes on the use of traditional highland Maya calendars to interpret the Codex Dresden, and the surviving portion of the original manuscript includes a discussion of eclipses and the eclipse tables identified in the Codex Dresden. This is followed by information from the indigenous communities of Aguacatan, Chajul, Chalchitan, Momostenango, Nebaj, and Todos Santos on the astrological and other attributes for the numbers and day names used in divinatory calendars. Burkitt's notes are especially important for his attempt to identify and define variability in the significance of aspects of the traditional highland Maya calendar. There are no dates of composition, although on page 3601 Burkitt indicates that some data were collected during the period from 1918 to 1923.[1]

The initial sections of Burkitt's manuscript have been lost. The existing text begins with section 5 of his original manuscript.

---

5. The modern Indians have lost not only (of course) the great year, but the period of 52 years, and even (yet they say) the quadriennium. But the period of 260 days survives, and in some places the year, the common (which is also the Indian year) of 365 days. Now in the neighborhood of K'iche' and Chajul, and in the Mam country, the Year Bearers, like those of the pot, and of the Atlantic monuments, are [E], [No'j], [Iq'], and [Keej]. And according to the existing calendar, it turns out that the last time that a new year of [5 E] fell at midsummer was in 1523; as you can evidence.

6. In the year 1900, by that K'iche' system, the Indian New Year was the 21st of March,[2] and was [5 No'j]. The 1523 New Year was 377 years earlier; 377 × 365 days. 377 happens to be divisible by 13 without remainder; so for the magic number, the 1523 New Year must bring back the number 5. What was the name? In 1524 (divisible by 4) the Indian new year was evidently [5 No'j] as in 1900. In 1523 it was consequently [E]. The 1523 New Year was [5 E].

7. What day of our year was it? Take Julian numbers:

1900 March 21	=	2415100
377 × 365	=	137605
1523 June 11	=	2277495

The 1523 New Year of [5 E] was June 11; in new style, June 21, i.e., midsummer.

8. What was the precise midsummer moment? The almanac that I've got hold of is for 1910. 1523 is 389 years earlier. From midsummer 1523 to that of 1910 will be 387 solar years. 387 × 365.2422 = 141348.731 in which I suppose there may be an error of 387 × ±0.00005 (−0.000001), or ±0.02. Neglecting that error, amounting to ± half an hour (less than a minute), subtract the 141,348.731 from the Julian number and fraction, for the midsummer of 1950 . . . 2418844.826 and 1523 June 12, 995 = 2,277,496.095.

The moment of midsummer was not on June 11, but a little after Greenwich noon on June 12, so that if that new year of [5 E] was expected at the end of 1,500 years to come exact on midsummer, the Indian clock (so to speak) turned out about a day fast. Taking the point of noon, of that New Year, June 11, noon, at 90W, was Greenwich June 11.25; midsummer exact was Greenwich June 12.095, or a difference of 0.845.

The Indian clock was fast by 0.845 of a day, or say 20.5 hours, with the exact midsummer about half past eight on the morning of the next day, the day of [6 Aj].

9. And there it may strike you that in Solomá and these northwestern places, with their new years always the tomorrows of those of the southeast, it may strike you that that day of [6 Aj] was precisely the Solomá new year; so that for a new year that might fall at midsummer, and so be appropriate for the start or finish of a great year; that Solomá new year of [6 Aj] would have been more exact than the pots [5 E]. There's no sign that that [6 Aj], even in Solomá, was ever so used; but signs are not wanting that in one way or other the pots [5 E] had competition.

10. How easily it might have had competitors, even within its own quartette of Year Bearers, you may see in [Table A1.1].

Not only, peruse, in 1523, but year after year, in that age, the Indian New Year fell also to midsummer. Seven hundred and fifty years earlier it had been falling as far as possible from midsummer. It had been falling at midwinter. By the year 1500, the difference of six months had come down to less than five days. In 1504, it was three days and three-quarters. In 1507, it was just three days. In 1519 and [15]20, it was at length practically nothing; the New Year's Day and the midsummer's day by our calculations, as nearly as possible, coinciding. In 1534 the difference is again just three days, as in 1507; but with the sign of what the table calls Slow turned into what it calls Fast. The New Year's Day, instead of coming after midsummer, comes before. It's in retreat. It has started again on its long journey to midwinter.

Table A1.1. Correspondence of year bearers and calendar dates.

AD	From Noon Midsummer (Greenwich)	Indian New Year	Noon of INY at 90' W	Days Fast (F) or Slow (S)
1492	June 11.59	13 No'j	June 18.25	S6.66
1500	June 11.52	8 No'j	June 16.25	S4.73
1504	June 11.49	12 No'j	June 15.25	S3.76
1507	June 12.22	2 E	June 15.25	S3.03
1511	June 12.19	6 E	June 14.25	S2.06
1512	June 11.43	7 No'j	June 13.25	S1.82
1513	June 11.67	8 Iq'	June 13.25	S1.58
1514	June 11.92	9 Keej	June 13.25	S1.33
1515	June 12.16	10 E	June 13.25	S1.09
1516	June 11.40	11 No'j	June 12.25	S0.85
1517	June 11.64	12 Iq'	June 12.25	S0.61
1518	June 11.88	13 Keej	June 12.25	S0.37
1519	June 12.13	1 E	June 12.25	S0.12
1520	June 11.37	2 No'j	June 11.25	F0.12
1521	June 11.61	3 Iq'	June 11.25	F0.36
1522	June 11.85	4 Keej	June 11.25	F0.60
1523	June 12.10	5 E	June 11.25	F0.85
1524	June 11.34	6 No'j	June 10.25	F1.09
1525	June 11.58	7 Iq'	June 10.25	F1.33
1526	June 11.82	8 Keej	June 10.25	F1.57
1527	June 12.06	9 E	June 10.25	F1.81
1528	June 11.30	10 No'j	June 9.25	F2.05
1529	June 11.55	11 Iq'	June 9.25	F2.30
1530	June 11.79	12 Keej	June 9.25	F2.54
1531	June 12.03	13 E	June 9.25	F2.78
1532	June 11.27	1 No'j	June 8.25	F3.02

11. The ancient Indians, or the astronomers among them, no doubt saw the same fact, at least the broad fact; but saw it perhaps only broadly; certainly not with the precision that you see it in the table. In the table, taking as I've done, the existing K'iche' succession of new year days, falling forever on the four names of [E], [No'j], [Iq'], and [Keej], and supposing too that the Indian standard moment of observation was the moment of noon; you see at one that your choice for a new year day that may coincide as nearly as possible with midsummer is limited to [a] choice between the two years, that is said, of 1519 and 1520; between the 1519 new year of [1 E], and the 1520 new year of [2 No'j]. But it's not likely that the early Indian astronomers so saw.

12. By observations covering many years, or theoretically, by no more than two good observations, provided they were many years apart, the Indians might early

know the length of the tropical year (or as they would perhaps have considered it, the annual celebration of the sun) with decided accuracy. For the sake of argument, let them know it, as well as our own astronomers. And yet, depending on their times and methods of observation (of which practically nothing is known) not to speak of chance accidents such as weather [*sic*]. You can imagine that they might not be certain of the precise moment of midsummer within anything closer than perhaps some three or four days (other accidents too, as our frequent St. Johns Day for midsummer).

The discovery of the New Year coinciding with midsummer would lie under the same uncertainty. Slipping my table back to the distant age when the question, in whatever way, first came up, you see that the wise men of one town or kingdom might take up with a year answering to my 1507, considering as midsummer (or perhaps midwinter) and adopting as their beginning of years the then new year of [2 E]. Twenty-five years later, in the year answering my 1531, another people in the same way, and equally three days astray, might adopt this new year of [1 No'j]. In the intervening years, according, perhaps, as the fashion traveled, other people might adopt intervening New Year days, their fundamental supposed midsummer (or midwinter). The choice of the appropriate New Year, instead of being plainly limited to the two years answering to my 1519 and [15]20, would be a matter of dispute, and seems to float uncertainly among all the ten or twelve, or even more years, on either side of that point. Each people would stick obstinately to their own choice. The chance new year, that they had raised to the dignity of making the starting point of their new establishment, not only was a better starting point of their new first year, but by the same fact, the starting point of their new first quadriennium, and of their new first 52-year cycle, and finally, of their great year. At the end of 1,500 natural years, the chosen day would return, and again a new year, again close to midsummer (or midwinter), with the same little error as when it set out. And so would all its competitors return, each in its place; and your words have the spectacle of what you might call a crop of great years all maturing together; that is, within a distance of a few years. To be sure, some competitors might drop out. Original uncertainties about midsummer (or midwinter) might vanish. Corrections might end in assimilations; or competitors might be extinguished by conquest. But you see how easily, whether on grounds of poor astronomy or of politics, or religion, or of mere blunders, or of all combined, how easily such a competition as I speak of is imaginable. And now as I say, evidence as not wanting that, the imaginable competition has existed. The fundamental new year of 5 Tooth [5 E], which in the age of the conquest came to stand about midsummer, there's a scattering of living evidence that it didn't stand alone.

13. Memoranda. All Saints [Todos Santos]. Numbers unknown. Chiantla, preeminence of Magistrat day [E]; [B'atz'], bad [a step toward Chuh?].

14. Chajul. Magistrat days respected in all numbers, 1–13, but the blessed numbers are 2, 4, and 8; with 8 the chief. These are also the leap-day numbers. Less, but

valued, are 6, 10, 13. For wicked or counterwork, 9, 11, 12, and even litanies of 9's, and of 13's, or of 9's and 13's together. A litany of peculiar complication. Notion of even numbers. Possibly mistaken. Remember that Diego Cara and Jacinto Velasquez are known as instructors; Cara one of the Council of Ten.

15. Days. [Q'anil] and [Ajmaq] (both females) are for corn. Corn only; for all crops. [B'atz'], for women and childbirth. Magistrats: they conclude doubtfully that the fierce are [Iq'] and [Kiej], bring rain and wind; not fierce, [E] and [No'j]. Remarkably favorable mention of [No'j].

16. Nebaj. 2908. Jacinto Velasquez, instructor of doctors. The blessed numbers are 5, 7, and 8. He constantly refers to them, even on bad days. Note to presenter, to burn, on those numbers would be punishable. He will not burn on the coming [12 Junajpu], he will burn on the following 6. But again he says put down 7, 8, 5, only the three. 2921. The 8, the 7, and the 5, mentioned as numbers for firing rockets (cf. the Chajul leap numbers). He seems to make no difference in value among the good numbers; the 8 no better than the 5, and so on. And no doubt the formal prayers don't vary. Yet a superiority of esteem for the 8 appears in the case of the bad day [Aq'ab'al]. For asking favors of that day (which is what is in his mind) he likes no number but 8. 8, yes, to beg that your cattle may not be bitten by the bat; or to beg leave, or pardon for stealing. So with other people, a man of 60, though not a professional burner, says that the nicest numbers are 8 and 7, the 8 best of all; a younger man (D.S.) but constant burner (like his father before him) calls the good numbers 1–8, especially 8.

17. Less than the blessed three, but, well esteemed, with J[acinto] V[elasquez], at least for Magistrat days, and other fair days, are the numbers 6 and 10, as in Chajul; and thirdly, not the Chujul 13 (see below) but 11. As for the low numbers, 1/4, they may sometimes be useful but he evidently doesn't much value them. Case of 4 taken but not 1, 3045.

18. At the opposite pole from 5, 7, and 8 are 9, 12, and 13. It seems that he may respect Magistrat days, even in those numbers (case of [9 No'j]); but constantly, no matter what the day, he mentions those three numbers only to reject them. 2917: he will burn for a coming [5 Kawoq], but not for the following 12. It would be a sin. So also the 13. No burning on [9 Tijax]; unless for counterwork. He will burn on the coming [6 Kiej], but not on the 13 or 12 (case of [9 Iq'] rejected), and so on. Evidently 9, 12, and 13 are peculiarly the numbers for wicked "work," wicked prayers; or with good people (like him) for counter-prayers.

19. The simple notion of Diego Brito[3] and others that even numbers are good, and odd numbers bad, J[acinto] V[elásquez] merely laughs at. What does Diego Brito know? But that's not to say that he would laugh at any popular statement. It would seem to be a popular Nebaj statement that the good numbers are from 1 to 8, and the bad from 9 to 13. In a statement that, no doubt, expresses broadly the mind of the town (as it might almost equally express the mind of Chajul) and the doctor

would no doubt in some way allow it. But subject to explanation. As for the 1/8 being good, there is no question (he would say) about the 5/8. And the 1/4 may be called good, as a consequence of their weakness. They are too weak to be used for anything bad. Again, the lumping of the 9 to 13 as bad, is a consequence of their strength. The 10 and 11, that our religion, at least with fair days, points out as suitable for good work, for asking blessings, may no doubt, whether fair day or foul, be laid hold of by bad people (and by good people to counteract the bad) in the same way as the 9, 12, and 13, and the whole bunch gets a bad name.

20. A couple of the half-instructed (a Michael Raymundo, caporal of S. Gordillo, with a G. Zedillo, said to know all about the 20 days) went so far as to put even the number 8 among the numbers too strong for honest "work"; as I suppose it might conceivably be, for a timorous burner, who felt himself inexpert, or had something on his conscience. But they curiously pointed out one little thing that the doctor would have quite agreed to; a certain alteration of good and bad. Today, 5 Aáma, they say, is good. But the next Aáma will be 12, and bad; while the next again will be 6, and good. Tomorrow will be [6 No'j], and good. Twenty days later turns up [13 No'j], not good; a day for bad people, or to work against them. And, of course, the successor again of 13 is 7; which even with them, is good. But the alternation of good and bad in that sequence, 5, 12, 6, 13, 7 (which was as far forward as they looked) would then fail; the next number 1, being (in their way of speaking) good.

21. Days. [Q'anil] and [Ajmaq], females, and for crops. [B'atz'], women and childbirth; Bird, for [B'atz'], etc., as in Chajul. Magistrats. No Magistrat in the 5 days (1987). Nothing like the Momostenango Second Mam, Magistrat days the birthdays of doctors. [Iq'] and [E] likely to bring rain and bad crops. [Kiej] and [No'j] make good (M.R. of no. 20); [No'j] was then late coming and is an example of a set of burnings on four successive Magistrat days, which happen to begin with [No'j] (V.Z. of no. 16). It was then the tail end of a [Iq'] year. What Magistrats are fierce? Deben de ser iguales. Dicen que Iq es bravo, . . . a la milpa el viento. Also, [No'j] rains a great deal they say. But he goes on to say that [Kiej] and [E] are also fierce. 1982, Diego Soto, with his father Peter, cf. no. 16, calls [No'j] "mas milagro que los demás alcaldes," the same Diego: the best days of all are the four Magistrat days. And the greatest day in [No'j], [8 No'j]. 2932, the doctor J. V. of no. 16 (it was then on Tooth [E] year); to ask to have the clearing burn well, the days are not [Q'anil] and [Ajmaq], but [Iq'] and [E]; those two, not the other Magistrats [Kiej] or [No'j]. A [Iq'] year is apt to be windy, and a [E] year a year of much hail. He had just been saying that the days to pray for rain or fine weather were not [Q'anil] nor [Ajmaq] (the crop days) but the four Magistrat days.

So setting aside the V.Z. of 1963, who plainly has no opinion, the doctor and another profess an opinion that the fierce Magistrats are [Iq'] and [E], the kind [Kiej] and [No'j]. All towns that preserve these four year bearers call [Iq'] fierce.

The question is, what shall be its companion? [Kiej], or [E], or [No'j]? Nebaj and Chajul seem to agree making it not [No'j]. They both seem to make [No'j] kind. And the statements of Diego Soto are striking; Earthquake [No'j] more wondrous (in answer to prayers) than the other Magistrats; and [8 No'j] the greatest of all days. Might [No'j], in the Ixil towns, have been the fundamental new year? By the name of the day, No'j, seems to have no meaning in the language.

22. Litanies. 20 days called on. 260 days called on. But as for the Chajul litanies, Diego Soto and father remark on them only to wonder and to misunderstand.

23. [Nebaj]. Work of 20 days running. No hills or burning places associated with particular days or numbers. Weeks by name, times of the year, are (both here and in Chajul) Amaq. 260 day round, 13 × 20. Year = 18 × 20 + 5.

24. Aguacatan, like Chajul, and Coban, is a coalition of originally distinct bodies. The creoles divide the town into Aguacatan and Chalchitan; both of which Mexican names, besides Balimajai and other native names, are set down by Remesal in his mention of the coalescing villages, among which by mistake he once includes even Nebaj. Balim or Balimaja, indifferently, is at this day the Indian name of the town of Aguacatan; both in Aguacatan itself and in Chiquimula, Momostenango. In Cunen, Balimja, and Balamja; in Uspantan, Boolomja in the town of Balanya, near Tecpan, is Balanyan and means Leopard Water. But that can hardly be the meaning of these names of Aguacatan, which should then appear as Balimajan, and so on. The Cunen Balamja seems to say plainly Leopard House, although not dwelling house, and Balimaja may be a lost dialect. House in Aguacatan is in all senses ká^al; while cat beast is báalum, or báluum. In no language that I know, is it balím. It has to be said, though, that the name of Balim or Balimajá is limited, with the Aguacatan people themselves, to the eastern or lower half of the town, the half toward the Ixil and Pichiquil; and that the Ixil and Mam names of the town have no resemblance to Balím or Balimajá. The Ixtahuacan and Chiantla names, Qtxíjel, Qtxíyel, and the Nebaj and Chajul name, Tsclúl, in some combinations Tsel, represent no doubt the names of some other of the original components of Aguacatan. However that may be, and small as is the country of the Aguacatan language, there are decided differences of dialect; as the people themselves will tell you. I've noted how the stiff *p*, preserved by the downtown or Pichiquil side, is turned into *b* by the uptown or Chiantla side. And with differences of dialect its possible other differences of superstition.

25. Numbers. My chief informants are A, B, C, D, E. A recommended by his nephew in Chocolá, and accompanied for a while by his partly drunk brother, is an uptown man who has been First Regidor of the town. They make out that today is [9 Toj] by reckoning from the last day [I'x], which was [7 I'x] 15 days ago; have they been so long drunk? The great numbers with them plainly are 2 and 7. [2 B'atz'] very merry (alegre), also [2 E]; and [7 B'atz']; not 8. 1, 3, and 4 of [I'x] are good, especially the 4, but not many, like the 2; the 2 goes with the 7, mui alegre.

26. B. Man of Pichiquil, about 45, employed by a labor agent, has been an under officer (mayor) of the church society (cofradia) and expects promotion. Today is [3 Aj]. 1/5 he calls weak, flojos. Yesterday (though the great day [E]) was [2 E], and consequently weak. The best numbers (he says) are 8, 12, 10 (his order). Not 7, at all. Afterwards however, he makes a good deal of 7, along with 5, 9, 12, and now and then a 10, for use on foul days or for counterwork. 11 and 13 he seldom mentions, perhaps leaving them mainly for bad people. For middling days he seems to fall back on the higher low numbers, 3, 4, 5, but still with 6 and 8 and now and then 10 or 12. For the great day [B'atz'], he gives 6, 8, 9, 12, with 8 as chief; and for [E] 6, 8, and not 0, but 12, with 8 as chief. He vacillates; but his most constant numbers, especially with fair days, are 6, 8, and 12; with 8 the chief. Plainly (though his practice may be less influenced than his talk) what the man has in his head, apart from weak and strong, is odd and even: 6, 8, 10, and 12, on the good side, and 5, 7, 9, 11, and 13, on the defensive, or bad.

27. C. Two uptown men, servants (mayores) at the courthouse, who like B, have been similar servants in the church society, but don't profess to know much, saying that only the diviners (zahorines) know well, seem to know particularly little about the numbers. For [B'atz'] the best numbers are not 1, 2, 3, which are weak, but 6, 8, 9, 12; with 8 and 9 the chief. For Tooth [E], 3, 4, and others, stopping at 8. They mention uncertainly one or two other 8's. But while seeming to support B in valuing 8, they seem to support A in once or twice valuing 2; for [No'j], 2; for [Junajpu], 2 (not 1).

28. D, an uptown man, about 50, a professional bean-counter, goes annually to the feast of Chajul (though not to the hill of Huil), and has the tricks of the trade. Was never in the church confraternity, and what he knows he got partly from his Indian teacher, but mostly from God. Today, he says at once correctly, is [8 Imox]. Everybody has prayer-making (costumbre) today; and it turns out that he has two clients for today. He lays it down that the good numbers are 5, 6, 7, 8; these four, 1, 2, 3, and 4, are too weak; those after 8, too strong. What is the next Monkey [B'atz'] finds correctly that it will be 5. Will that be a big day? Yes. And the following 12? No, but the subsequent 6, and the 7, and the 8; nothing beyond. No doubt however, he uses the higher numbers for counterwork; and if I had gone through the days by numbers, he would probably have found opportunities for one or more of the low numbers. As for his 5, 6, 7, and 8, there is no one of them that has more or less support among previous speakers. Even the 5 has some echo with B. But what's more remarkable is how those numbers of his, so far as they go, repeat the good numbers of Nebaj: 5, 7, 8 (no. 16); 6, 10, 11 (no. 17).

29. E. Man, say 40, never in the church fraternity, has been the employer of a labor agent, and was recommended to me by the men of C. Has been drunk and is still muzzy. I don't ask him much about numbers, but make out that for this same day [Imox] the good numbers are 8, and 1, and 2. He talks of [8 No'j], and also the

1 and 2. But later, he seems to disparage number 1. Volunteers for [Tz'i'], the number is 13, and so with other foul days; thinking no doubt of counterwork. What's the nice number for [B'atz']? 8. For [E]? 8. For [Kan]? 8. And won't have 1, or 2, or even 13 [Kan]. Only 8. Whatever he might add in another state of mind, it's clear his great number is 8.

30. On the whole, I should gather that the half dozen best numbers for honest prayers, in Aguacatan, were 2, 5, 6, 7, 8, and 12. To put numbers and speakers in a table [Table A1.2]:

Table A1.2. Numbers for honest prayers, Aguacatan.

2 is supported by A; slightly by C and E
5 is supported by D; slightly by B
6 is supported by B, C, D
7 is supported by A, D; slightly by B
8 is supported by B, C, D, E
12 is supported by B; slightly by C

31. The table doesn't show the disagreements. The man D, so far as he goes, ignores equally the 2 of A, and the 12 of B. With respect to the number 2, and even in the precise case of the important day [E], A and B flatly contradict each other. And, almost incredibly, the number 8, that all others agree on, A has the nerve to reject even in the case of [B'atz']. Such differences as these last; whether from doctor to doctor or from district to district, make it clear that the Aguacatan religion of the numbers is in the last stages of decay. The man D, in spite of his claim to inspiration, I take to be my chief informant. His set of 5, 6, 7, 8 is the most comprehensive of any and its likeness to Nebaj is not at all a miscommunication. I pass over the 9, mentioned A, B, and C, as merely an affair of [B'atz']; and make up my half dozen by taking all four numbers of D, and adding the two non-D that seem best supported. No doubt 8 was the chief of the Aguacatan blessed numbers. As for the next two, I can only guess that they were some two out of the present four favorable numbers below 8.

32. Days. Excepting with settlers from Chikimul [Chiquimula?], the year is lost. Not only are there no names of weeks but there is no notion of the five days, or the Magistrate days. A week as a period of 20 days is junwíng; two weeks, kxówíng; three weeks, ox wing; the element wing not being the usual Aguacatan for 20, but a special form echoing forms for 20 of Mam or Ixil. So Solomá says jun wináq to mean junk al kú, 20 days. As for the 260-day round, neither Aguacatan man that I asked could say how long it was; but they might know it in their own way.

33. In the Ixil towns the name of the Lizard [K'at] day, which is katx, might equally mean burn or net. In Aguacatan the sound is the same, katx, but the meaning of net is eliminated; net being kaatx. In Aguacatan and the Ixil towns alike its names of the Snake day is Kaan, and is not the word for snake; that word in the Ixil

towns being Kan, and in Aguacatan, lubáj. In Aguacatan, however, the name of the day has the meaning of pains, pains like cramps or rheumatism, another case, no doubt, like that of Lizard [K'at] day in Coatan used with the meaning of Bat and so on. The name of the day has become, by superstition, a byname for something else. The name of the Flint [Tijax] day, if the name (txi^j) is used to mean angry, may be another instance.

34. [Aq'ab'al], instead of foul, as in the Ixil towns, and Todos Santos, is now called fair. But still connected with night, to beg a wife, to beg information by dreams.

35. [Tz'i'] and [Kawoq], both foul, as in the Ixil towns and in Todos Santos, here put together once as patrons of fighting. Kaan and [I'x] put together against coyotes. Rabbit [Q'anil] and Vulture [Ajmaq], though no longer females, continue in the Ixil towns to be the twin patrons of corn and crops. They are also mentioned for money. In spite of which, each of them has some black mark against it, and may be called only middling fair; the mark against [Q'anil] (preserving in Aguacatan the superstition of Mexico) being that [Q'anil] is the patron of drunkenness. See Sahagún.

36. The Magistrate days being no longer Magistrates, anything may happen to them. [Iq'] is simply a foul day. [No'j], once called foul, is usually fair, among the fairest. [Keej], once called foul seems to be middling. It has perhaps a trace of its former Magistrate character in being once called a day of Wizards. See no. 21. But what is more curious about it, [Keej] is now paired with another day. It has become contaminated by the day before the day [Kame], and is paired with [Kame] as jailer of the dead. In Chajul (though not in Nebaj) a day that in the same way paired with [Kame] is the day before [Kame], the otherwise different and highly respected day [Kan]; a day that in Aguacatan is simply foul.

37. And another case in Aguacatan like that of [Keej], another case of contamination, may seem to be the case of the remaining ex-Magistrat day, the day [E]. The day before [E], the day [B'atz'], without having (as it has in the Ixil towns) any special ear to prayers of women, is yet in Aguacatan a general fair day, and not simply fair, but fair to such a degree that only one other day of the 20 seems to be its equal or superior; that other superlative day being the next day, this ex-Magistrat day of [E]. I was once in the town on a day [7 E]. The consequence of the day was obscured by the day's happening to coincide with an annual feast fixed to the Christian calendar. It's not likely, however, in that town, that even the best days and numbers cause any visible crowd in church or market (as happens in some places) or the creoles would know it. Yet there can be no doubt of the extreme estimation, in Aguacatan, of both [B'atz'] and [E].

38. Now the high esteem of [B'atz'], appearing again strongly in the towns of K'iche', is no doubt prehistoric. But what about [E]? Might the high esteem of that day be a more modern result of the loss of the year? A mere contamination of

the empty Magistrat day by the preceding [B'atz'], as [Keej] is contaminated by [Kame]? I suppose it might. But looking at the Mam religion, looking at All Saints [Todos Santos] I should guess that something else was more likely. In All Saints [Todos Santos], where the day [B'atz'] is reckoned foul, [E] is never the less pre-eminently fair, just as in Aguacatan; and in addition is still a Magistrat; one of the four new year days; the year, in All Saints [Todos Santos], not being lost. I should guess that the greatness of [E] in Aguacatan was one with its greatness in All Saints [Todos Santos], and that in both places Tooth [E], perhaps, had been the funda-mental new year.

39. Hills.

40. Momostenango. My response statistics of pp. 2394ff, belonging to the years 1918–23, cover about six rounds of 260 days; and the tabled amount for each Indian day and number is the mean of four or five or six returns of the day; depend-ing on the absence of the priest or other interruptions. The amounts there set down are money of the country, and nearly the same thing as to say so many pence; pence which are the churches' little dues. The 260-day table of p. 2394 shows a mean total of near 1,000 pence; signifying in responses (at a farthing a response) a total of near 4,000 responses. To make it handier, I now turn to that table [Table A1.3] showing the 260-day distribution of a mean total (of pence or responses) to the number of simply a thousand.

The blanks, as before, are naughts. In a few places, on last adjustment, the sup-pression of fractions has made me do a little squeezing, made me increase or dimin-ish the preliminary whole number by one unit. The mark of those places is a plus or a minus [before]; telling me (if I ever wish to know) that the number first calculated in the case of [4 No'j] (for example) was not naught, but one; and in the case of [8 No'j] was not 16 but 15. From that table I extract:

41. The thirteen numbers in the order of responses:

[Number]	8	1	6	9	7	5	11	13	2	3	4	12	10	All
[Responses]	449	268	140	79	16	12	9	7	5	4	4	4	3	1000

42. Twenty day names in order of response, Momostenango [Table A1.4].

In the case of [No'j] the original mean total was 354, and in the case of [Toj], 353. The difference of one, in near 10,000, is now advantageously sunk.

43. The 26 best days, in order of responses [Table A1.5].

44. A continuation of this list would be nothing but a smooth downhill, end-ing in the dead level of the 150 days or so that the table shows blank. In the list as it is, you see that a few best days, the best 13 as it happens, account for fully half the responses of the whole 13 weeks. And seeing that just at that point, between the 13th day and the 14th, there happens to be sudden break in responses; the last break of any consideration in them. You might think of taking that point as a point for dividing the days (with respect to responses) into greater and lesser; calling the

Table A1.3. Correspondence of calendar dates and church donations, Momostenango, 1918–1923.

	1	2	3	4	5	6	7	8	9	10	11	12	13	All
E	12	—	—	—	—	3	—	26	2	1	—	1	—	43
Aj	11	—	—	—	7	14	11	32	3	—	—	1	1	80
Iʼx	5	—	—	—	—	1	—	4	1	—	—	—	—	11
Tzʼikin	14	—	1	—	—	5	—	34	—	—	—	—	—	54
Ajmaq	12	—	—	+	—	11	—	16	2	—	1	—	1	42
Noʼj	6	—	—	—	—	9	—	16	3	—	—	—	—	36
Tijax	4	—	—	—	—	2	—	9	1	—	—	—	—	16
Kawoq	9	—	—	—	—	6	—	10	—	—	—	—	—	25
Junajpu	36	4	3	3	+2	16	3	33	6	—	4	2	1	113
Imox	3	—	—	—	1	3	—	3	—	—	—	—	—	10
Iqʼ	6	—	—	—	—	2	—	4	—	—	—	—	—	12
Aqʼabʼal	26	—	+	—	—	16	1	26	12	—	1	—	—	82
Kʼat	3	—	—	—	—	2	—	4	—	—	—	—	—	9
Kan	5	—	—	—	—	3	—	9	1	—	—	—	—	18
Kame	16	—	—	—	—	6	—	17	1	—	—	—	—	40
Keej	33	—	+	—	—	16	—	43	6	—	2	—	2	102
Qʼanil	18	1	—	1	1	9	1	27	—	2	1	1	1	63
Toj	14	—	—	—	—	5	—	61	+4	—	—	—	1	36
Tzʼiʼ	2	—	—	—	—	3	3	—	—	—	—	—	—	8
Bʼatzʼ	33	—	—	—	—	8	—	+122	37	—	—	—	—	200
All	268	5	4	4	12	140	19	446	79	3	9	4	7	1000

[Editor's note: This table is reproduced here as it appeared in the original document. It is likely that the "61" under Toj, row 8, should be "11," which would make the totals correct.]

Table A1.4. Eighteen day names in order of response, Momostenango.

Day Name	N	Day Name	N
B'atz'	200	Toj	36
Junajpu	113	Kawoq	25
Keej	102	Kan	18
Aq'ab'al	82	Tijax	16
Q'anil	63	Iq'	12
Tz'ikin	54	I'x	11
Ajmaq	42	Imox	10
Kame	40	K'at	9
No'j	36	Tz'i'	8
		Total	1,000

[Editor's note: This table is reproduced here as it appeared in the original document. Because E and Aj were left out of this table, the total is 877.]

Table A1.5. The twenty-six best calendar days in order of response, Momostenango.

Day	N	Day	N
8 B'atz'	122	1 Q'anil	18
8 Keej	43	8 Kame	17
9 B'atz'	37	6 Keej	16
1 Junajpu	36	6 Junajpu	16
8 Tz'ikin	34	8 Ajmaq	16
1 Keej	33	1 Kame	16
8 Junajpu	33	8 No'j	16
1 B'atz'	33	6 Aq'ab'al	16
8 Aj	32	6 Aj	14
8 Q'anil	27	1 Tz'ikin	14
1 Aq'ab'al	26	1 Toj	14
8 E	26	1 Ajmaq	12
8 Aq'ab'al	26	1 E	12
Best 13	508	Best 26	705

greater days the thirteen above that break (as in diagrams, etc.). But that last break is by no means the first. Nowhere really in the list is there any plain undeniable division, excepting the division between all other days and the top day; that one soaring day of 8 Monkey [8 B'atz'].

APPENDIX TWO

Notes on the Correlation of Maya and Gregorian Calendars

CHRISTIAN M. PRAGER AND FRAUKE SACHSE

In Calendar A of the 1722 K'iche' document, we find a number of calendar-round dates that are correlated with Gregorian dates. These dates provide a significant and hitherto unexamined argument in support of the Goodman-Martínez-Thompson (GMT) correlation (correlation constant = 584,283 days) of the Maya and European calendars (Thompson 1934).

The calendar correlation may be stated in the form of the "Ahau equation" (Willson 1924:16):

$$(1) \quad Day_{Mayan} + A_{Constant} = Day_{Julian}$$

Equation 1 describes a correlation of the Maya with the Julian day count. Since the Maya system of numeration uses positive natural numbers, the formula allows only integer solutions and the unit for the equation is "day." Calendar dates expressed in the Maya and Gregorian mode need to be converted into a common system of a continuous day count. Thus, the units Day_{Julian} and Day_{Mayan} describe the number of integer days reckoned from hypothetic zero days of the Gregorian (i.e., January 1, 4713 BC) and Maya calendar system respectively (i.e., 13.0.0.0.0 4 Ajaw 8 Kumk'u). $A_{Constant}$ is a numerical constant, or the quantity of days that must be added to or subtracted from Day_{Mayan} to give Day_{Julian} (Willson 1924:17).

Calendar A indicates dates in the form of calendar-round notations, which are correlated with absolute or unique dates indicated in the Gregorian calendar in the form of Julian dates:

(2)   December 9, 1722 (Julian) = 8 Kej 20 Ukab' Si'j

In order to solve the problem of the 1722 K'iche' calendar correlation, we follow Floyd Lounsbury, who explains: "The value of the left-hand member of this equation [. . .] is unique; that of the right-hand member [. . .] is not. Since such a calendar-round day recurs every 18,980 days, we should rewrite the left-hand member so as to make it similarly unspecific" (1992:187). Lounsbury suggests an expansion of his formula by adding $18,980k$, where k represents any positive or negative integer:

(3)   December 9, 1722 (Julian) + $18,980k$ = 8 Kej 20 Ukab' Si'j

Therefore, the correlation constant is not a specific numerical value, and one cannot determine the position within a linear time flow (Lounsbury 1992:185–187). Rather, the correlation constant must be determined mathematically within the length of a calendar round, a cycle of 18,980 days or fifty-two Vague Years.[1] This is illustrated in the following equation that is based on Equation 3 (Lounsbury 1992),

(4) Day_{Julian} + $18,980k$ = Calendar Round

where $18,980k$ is the expansion of the Julian day to make it as unspecific as the Maya calendar round and k is any integer.

Applying this formula to the correlation equation by substituting the variable Day_{Mayan},

(5) Day_{Mayan} + $A_{Constant}$ = Day_{Julian}

we may then calculate,

(6) Day_{Julian} − $A_{Constant}$ + $18,980\ k$ = n Tzolk'in m Haab

where $18,980k$ is the expansion of the Julian day to make it as unspecific as the Maya calendar round and k represents any integer. The date 'n Tzolk'in m Haab' represents the specific elements of the Maya calendar round where n represents the day numbers 1 to 13, and m the numbers of days of the Maya month.

This equation does not contain a numerical value for the correlation constant although it permits the substitution of known correlation constants as a check on their compatibility and plausibility. Any correlation constant for the 1722 K'iche' calendar can only be correct if the result of a calculation of *k* is an integer.

Lounsbury (1992) developed Equation 5 to calculate the correlation of the Classic Maya calendar with the Christian calendar. Before applying this equation to the 1722 K'iche' calendar, however, we need to establish whether the variable used in Lounsbury's equation for the Classic Maya calendar is the same as for the 1722 K'iche' calendar from the Guatemalan highlands. In order to find out, we need to examine whether names and sequence of the Tzolkin days and Haab months in both calendars correspond. This seems to be the case for the sequence of day names (La Farge 1934:112), but there are obvious differences in the sequence of months in the highland and lowland calendars. Table A2.1 compares the sequence of the months in the various sections of the 1722 K'iche' calendar, in the 1685 Kaqchikel calendar, and in the 365-day count of the Classic lowland Maya tradition. The numerals preceding the month names indicate the first and last month of each calendar.

La Farge (1934) and Thompson (1950:106) have identified variability in the sequence of months among Q'anjob'al, Pocomchi, Ixil, Tzeltal, and Tzotzil calendars. Similarity in the names of the months is no indication for the position of a specific month within the year count. The correlation of different calendars in Table A2.1 is not based on the similarity of names but instead on arithmetic calculation. This correlation demonstrates that the 1722 K'iche' Calendars A and C and the 1685 Kaqchikel calendar do not correspond in the generalized sequence of months.

The position of the twentieth day of a month in the K'iche' calendar corresponds with the numeral 0 as in the Petén style of dating, which begins the new year with the seating of Pop, or 0 Pop. However, 0 Pop is not to be correlated with the day of the month 20 Nab'e Mam, which is the first month listed in the 1722 K'iche' Calendar A. The K'iche' calendar and the Classic Maya calendar associate the beginning and the end of the year with different months. Pop corresponds with the K'iche' month Likinka. A comparison of the sequence of months in the 1685 Kaqchikel and 1722 K'iche' calendars shows deviation in the position of the five closing days (Tz'api Q'ij). In the 1722 K'iche' calendar the position of the Wayeb period is between the months Ch'ab and Nab'e Mam, which corresponds to Pax and K'ayab in the Classic period lowland tradition, but the 1685 Kaqchikel calendar places the Wayeb between Pariche' and Tacaxepual, that is, between Sak and Keh.

The variation in new year and the Wayeb day positions complicates any correlation of the Classic Maya and Christian calendars using the 1722 K'iche' calendar. A correlation based on the similarity of month names alone is not possible. The following calculation of the correlation constant for the 1722 K'iche' calendar considers that the sequence of months will vary in regional calendrical sequences and thus includes only the full *tzolk'in* date and the coefficient of the month, but not the month itself.

Table A2.1. Correspondence of calendrical month names.

	Classic period Maya	K'iche' 1722 A-I ff. 1–23	K'iche' 1722 A-II	K'iche' 1722 A-III-VI ff.	K'iche' 1722 C f.50	Kaqchikel 1685
1	Kayab	Nabe' Mam	Nabe' Mam	Nabe' Mam	Mam	Nabe'y Mam
2	Cumuh	Ukab' Mam	Ukab' Mam	Ukab' Mam	Ukab' Mam	Rukab' Mam
3	Pop	Likinka	Likinka	Nabe Likinka	Likinka	Likinka
4	Uo	Ukab' Likinka	Ukab' Likinka	Ukab' Likinka	Ukab' Likinka	Nabe'y Toqik
5	Zip	Nabe Pach	Nabe Pach	Nabe Pach	Pach	Rukab' Toqik
6	Zotz	Ukab' Pach	Ukab' Pach	Ukab' Pach	Ukab' Pach	Nabe'y Pach
7	Zec	Tz'isi Laqam	Tz'isi Laqam	Tz'isi Laqam	Tz'isi Laqam	Rukan Pach
8	Xul	Tz'ikin Q'ij	Tz'ikin Q'ij	Tz'ikin Q'ij	Tz'ikin Q'ij	Tz'ikin Q'ij
9	Yaxkin	Kaqam	Kaqam	Kaqam	Kaqam	Kaqan
10	Mol	B'otam	B'otam	B'otam	B'otam	Ib'ota'
11	Chen	Nabe Si'j	Nabe Si'j	Nabe Si'j	Si'j	Qatik
12	Yax	Ukab' Si'j	Ukab' Si'j	Ukab' Si'j	Ukab' Si'j	Iskal
13	Sak	Rox Si'j	Rox Si'j	Rox Si'j	Urox Si'j	Pariche'
				[Tz'api Q'ij]	[Tz'api Q'ij]	[Tz'api Q'ij]
14	Uayeb	Che'	Che'	Che'	Che'	Tacaxepual
15	Keh	Tekexepoal	Tekexepoal	Tekexepwal	Tequexepoal	Nabe' Tumuxux
16	Mac	Tz'ib'a Pop	Tz'ib'a Pop	Tz'ib'a Pop	Tz'ib'a Pop	Rukab' Tumuxux
17	Kankin	Saq	Saq	Saq	Saq	Q'ib'ixix
18	Muan	Ch'ab'	Ch'ab'	Ch'ab'	Ch'ab'	Uchum
19	Pax	[K'isb'al Rech]	Tz'api Q'ij	Tz'api Q'ij		

Analysis of the 1722 K'iche' calendar begins with the correlation of calendar-round positions with Gregorian dates. Twenty of the 380 correlations are overlaps that may be compared to indicate coherence. Table A2.2 gives the overlaps of Calendar A-I and A-II, which serve as the basis for a correlation. Columns one and five indicate pagination; columns two, three, six, and seven, the calendar round and corresponding date in the Gregorian calendar. The numerical values for each of the Gregorian dates (i.e., the days that have passed since January 1, 4713 BC) are given in columns four and eight.

One must calculate:

December 9, 1722 = 8 Kej 20 Ukab' Si'j

December 9, 1723 = 9 E 20 Ukab' Si'j

Inserting this information into Equation 5:

$$\text{Day}_{\text{Julian}} - A_{\text{Constant}} + 18{,}980\ k = n\ \text{Tzolk'in}\ m\ \text{Haab}$$

yields:

$$2350350 - A_{\text{Constant}} + 18{,}980\ k = 8\ \text{Kej 20 Ukab' Si'j} = 8\ \text{Kej 0 Ukab' Si'j}$$

where 2,350,350 is the number of days that have passed since the zero day of the Julian calendar.

Thompson (1971) and Schalley (2000) provide a summary of the various correlation constants (A_{Constant}) suggested in the literature. The numerical value for each of these has been examined to determine whether correlation in the 1722 K'iche' calendar is appropriate.

In the following calculation the day name in the form of n Tzolk'in and the day of the month are set quantities, as the correlation of month names with the months of other Maya calendars may give ambiguous results.

Use of the constants proposed by Thompson (1971), Schulz Friedmann (1955), or Kelley (1976) results in:

Thompson $2350350 - 584283 + 18980\ k =$ **8 Kej 0** "month"

Schulz $2350350 - 677723 + 18980\ k =$ **8 Kej 0** "month"

Kelley $2350350 - 553279 + 18980\ k =$ **8 Kej 0** "month"

Continuing the calculation further with these three examples gives the following results:

Thompson   December 9, 1722 = JD 2350350 = **8 Manik' 0** Yax [12.5.5.13.7]

Schulz   December 9, 1722 = JD 2350350 = **12 Manik' 0** Yax [11.12.6.3.7]

Kelley   December 9, 1722 = JD 2350350 = 7 Chuen 10 Mol [12.9.11.15.11]

Any correlation constant can be reasonable only when there are correspondences with all of the given values from the document. Using Thompson's correlation constant of A = 584,283, all values from the calendar round correspond with the information in the 1722 K'iche' calendar. With the correlation constant of A = 677,723 proposed by Schulz only two of the three values correspond, and Kelley's correlation constant does not correspond with any of the values in the 1722 K'iche' calendar. Similarly, only the values proposed by Spinden (489,383), Bowditch (394,483), and Thompson (584,283) correspond with the calendar rounds in the 1722 K'iche' document. Of these, however, only Thompson's correlation constant is acceptable in historical, astronomical, and archaeological contexts.

It is not possible to determine a numerical value for the correlation constant that is valid for the 1722 K'iche' calendar solely on the basis of correlating a linear time flow (Julian calendar number) with a cycle of 18,980 days and its multiple.

Any correlation constant that is applied in the equation

(1) $Date_{Maya} + A_{Constant} = Day_{Julian}$

must meet the condition expressed in the constant formula (Equation 6)

(6) $A = 14{,}883 + 18{,}980k$

in order to yield the correct calendar-round date indicated in the 1722 K'iche' calendar. This is indicated by the correlation constants proposed by Thompson, Bowditch, and Spinden:

Thompson $A = 14{,}883 + 18{,}980 \times 30 = 584{,}283$
Spinden $A = 14{,}883 + 18{,}980 \times 25 = 489{,}383$
Bowditch $A = 14{,}883 + 18{,}980 \times 20 = 394{,}483$

where k is always a positive integer. The equation serves to determine whether any other constants for the correlation of the Gregorian and K'iche' calendars are possible. Table A2.3 indicates only three of the correlation constants are reasonable because none of the others provide an integer from the equation.

This analysis demonstrates that only three of the correlation constants provided in the literature are acceptable with the dates given in the 1722 K'iche' calendar. At least one of the calendars provides, in addition to Lounsbury's (1992) Landa Equation, evidence for the testing of correlation constants. The calendar rounds as presented in the K'iche' document do not necessarily confirm that Thompson's ($A_{Constant} = 584{,}283$) correlation is correct, as there exists other evidence for this, but the result provides an additional argument. Based on the K'iche' manuscript alone, the arguments of Bowditch (1901) and Spinden (1924) are equally reasonable.

Thompson's ($A_{Constant} = 584{,}283$) correlation suggests that thirty-three of thirty-eight correlations from the 1722 K'iche' calendar (see Table A2.2) show correspondences in the day number, the day name, and the coefficient for the month.

Table A2.2. Basis for the calendrical correlation.

I	II	III	IV	V	VI	VII	VIII
Page	A-I (1722–23)	I	JD	Page	A-II (1723–24)	II	JD
1	9 Kej 20 Nabʼe Mam	May 3, 1722	2350130	18	10 E 20 Nabʼe Mam	May 3, 1723	2350495
2	3 Kej 20 Nabʼe Mam	May 23, 1722	2350150	18	4 E 20 Ukabʼ Mam	May 23, 1723	2350515
3	10 Kej 20 Likinka	June 12, 1722	2350170	18	11 E 20 Likinka	June 12, 1723	2350535
4	4 Kej 20 Likinka	July 2, 1722	2350190	18	5 E 20 Ukabʼ Likinka	July 2, 1723	2350555
5	11 Kej 20 Pach	July 22, 1722	2350210	18	12 E 20 Nabʼe Pach	July 22, 1723	2350575
5–6	5 Kej 20 Pach	August 11, 1722	2350230	18	6 E 20 Ukabʼ Pach	August 11, 1723	2350595
6	12 Kej 20 Tzʼisi Laqam	August 31, 1722	2350250	18	13 E 20 Tzʼisi Laqam	August 31, 1723	2350615
7	9 Kej 20 Tzʼikin Qʼij	September 30, 1722	2350270	18	7 E 20 Tzʼikin Qʼij	September 20, 1723	2350635
8	13 Kej 20 Kaqam	October 10, 1722	2350290	18	1 E 20 Kaqam	October 10, 1723	2350655

9	7 Kej 20 B'otam	October 30, 1722	2350310	18	8 E 20 B'otam	October 30, 1723	2350675
10	1 Kej 20 Nabe Sij	November 19, 1722	2350330	18	2 E 20 Nabe Sij	November 19, 1723	2350695
11	8 Kej 20 Ukab' Sij	December 9, 1722	2350350	18	9 E 20 Ukab' Sij	December 9, 1723	2350715
12	2 Kej 20 Rox Sij	December 29, 1722	2350370	18	3 E 20 Rox Sij	December 29, 1723	2350735
13	9 Kej 20 Che'	January 18, 1723	2350390	18	10 E 20 Che'	January 18, 1724	2350755
14	3 Kej 20 Tekexepual	February 4, 1723	2350410	18	4 E 20 Tekexepual	February 4, 1724	2350775
14–15	10 Kej 20 Tz'ib'a Pop	February 27, 1723	2350430	18	11 E 20 Tz'ib'a Pop	February 27, 1724	2350795
15	4 Kej 20 Saq	March 19, 1723	2350450	18	5 E 20 Saq	March 18, 1724	2350815
16	11 Kej 20 Ch'ab'	April 8, 1723	2350470	18	12 E 0 Ch'ab'	April 7, 1724	2350835
17	5 Kej 20 K'isbal Rech	April 28, 1723	2350490	18			

Table A2.3. Correlation constants suggested in the literature based on Fuls (2004) and Schalley (2000).

	Correlation constant	Value for k
Bowditch (1901)	394,483	20,0000000
Smiley (1961)	482,699	24,6478398
Owen (1977)	487,410	24,8960485
Spinden (1924)	489,383	25,0000000
Ludendorff (1930)	489,484	25,0053214
Teeple (1931)	492,622	25,1706533
Suchtelen (1958)	583,919	29,9808219
Goodman (1897)	584,280	29,9998419
Martínez (1932)	584,281	29,9998946
Thompson (1950)	584,283	30,0000000
Beyer (1933)	584,284	30,0000527
Thompson (1935)	584,285	30,0001054
Calderon (1966)	584,314	30,0016333
Schove (1982)	594,250	30,5251317
Schove (1982)	615,824	31,6618019
Fuls (2004)	660,208	34,0002634
Kelley (1983)	663,310	34,1636986
Dittrich (1936)	698,164	36,0000527
Weitzil (1945)	774,078	39,9997366
Vollamaere (1991)	774,080	39,9998419

Table A2.4. Deviations in the month coefficients.

1	9 Kej 20 Nab'e Mam	May 3, 1722	2350130	9 Manik' 5 K'ayab	12.5.5.2.7
2	3 Kej 20 Ukab' Mam	May 23, 1722	2350150	3 Manik' 5 Kumk'u	12.5.5.15.7
3	10 E 20 Nab'e Mam	May 3, 1723	2350495	10 Eb 5 K'ayab	12.5.5.2.7
4	4 E 20 Ukab' Mam	May 23, 1723	2350515	4 Eb 5 Kumk'u	12.5.5.15.7

The four examples in Table A2.4 show deviations in the coefficient of the months, indicating that the sequence of month names in the Maya highlands has shifted (Thompson 1950:310). This shift has affected the position of the new year and the five closing days.

Thompson's observation of the shift in month names in the Ixil calendar also holds true for the K'iche' calendar. Comparing the beginning point of the K'iche' calendar with that of the Classic lowland Maya calendar results in a difference of forty days or two K'iche' months:

May 3, 1722	9 Kej 0 Nab'e Mam = 9 Manik' 5 K'ayab	12.5.5.2.7
June 12, 1722	10 Kej 0 Likinka = 10 Manik' 0 Pop	12.5.5.4.7
	Difference	2.0
	(40 days)	

APPENDIX THREE

Agricultural Cycle and the K'iche'an Calendar

Maize (*Zea mays*; *ixiim*), a cultivated grass, is the principal food for the indigenous population of Guatemala. Various ethnographic studies (Carter 1969:14; McBryde 1947:16–26; Stadelman 1940:93; Steinberg and Taylor 2002; Tax 1953; Wilson 1972:139–141) have demonstrated that maize constitutes the greatest part of the Mayan diet, and is often almost the exclusive component. Beans (*Phaseolus* spp.), squash (*Cucurbita* spp.), various varieties of chili or peppers (*Capsicum* spp.), greens in season, and, rarely, meat, potatoes, eggs, and coffee are occasionally added. The usual daily ration of maize is more than sufficient to supply the daily energy requirements.

The crops typically planted with maize are beans (*quinak*) and chilacayote (*coc*; *Cucurbita ficifolia*), a type of squash similar to a greenish-white watermelon. Sometimes peppers (*yk*), tomatoes (*xkoya'*), potatoes (*yz*), squash (guisguil; *Sechium edule* or *Chayota edulis*), and inedible gourds for utensils are planted as well, according to local preferences. The beans are of two main types, the large kidney or butter bean (*Phaseolus coccineus*), usually called *piloy* or *chamborote*, and the small black or red bean (*Phaseolus vulgaris* L.). There are two methods of planting beans: the beans are planted either in the same hole with the corn or between the crops of maize. From two to four seeds are planted for each mat, and the growing plants send out runners that spread over the field.

Given the importance of maize to the Maya people of Guatemala and to their ancestors, it is not surprising that the cultivation of maize possesses great religious significance and is attended with elaborate ritual.

The scheduling of various agricultural tasks and the observance of maize ritual was traditionally selected by the daykeepers who were, in turn, governed by calendars such as those included in this volume. These calendars are ancient retentions, and certain day names, especially when they occur with certain numbers, are considered especially favorable for maize. Maize ritual always took place on one of these days. For Chichicastenango, Bunzel (1952:55, 282; see also Schultze Jena 1946:35 and Tedlock 1992:114 for Momostenango) notes that Q'anil is the day sacred to maize and the maize field, or milpa, and most closely associated with agriculture. The day 8 Q'anel is for the celebration of the fruitfulness of the earth and a day of obligation in the performance of ritual.

Unfortunately, these rituals are observed in only a few K'iche'an communities today. The indigenous population of Guatemala is sufficiently involved in the dominant Ladino culture to follow the European calendar in determining the time for planting.

AGRICULTURAL CYCLE

Information contained in the 1685 Kaqchikel and 1722 K'iche' calendars allows a concordance of the annual agricultural cycle (see Table A3.1 for a correspondence of the agricultural cycle and the 1723 calendar in the 1722 K'iche' calendar). The dates indicated in the calendars do not correspond precisely with ethnographic reporting. This, however, may be the result of scribal error or, more likely, from variation in planting patterns between communities because of differences in location, elevation, rainfall, and so on.

The agricultural year can be summarized by its essential features. Maize is grown only during the rainy season. The growing season is about eight months, extending generally from May through December, including the six months of the rainy season and the first two months or so of the dry season, when the ears of corn ripen.

Trees and bush on new lands are cut away in March and early April, and fields are burned during the last of April, when the milpa is ready for planting. Cutting and burning must occur before the first rains begin, the best burns occurring at the end of April or early May. Rain is a certainty by the end of May. Clearing involves pulling up the stalks of the last crop and hoeing under the weeds from the fallow season of mid-November to early April. The main ceremonial occasion associated with milpa comes at the beginning of clearing.

Planting occupies the main of the dry season, from February to mid-May, with firing planned on the eve of planting, if possible. Weather observations show the period April 2 to May 11 as the main period for firing. Planting in each area is completed in as few days as possible so that everyone's crop will reach each stage of maturity at nearly the same time. The reason is practical since all the hungry mice, rats, squirrels, raccoons, and birds will descend on the milpa of that man whose maize is out of phase.[1] Variation in elevation and temperature are the major obstacles to uniform planting.

Maize is planted after the first rains of May. Seeds that do not sprout are replaced. The milpa is cultivated at least twice or possibly three times. When plants are about a meter high the field is weeded and soil around each plant is hilled. Then a second weeding is done when the "points" begin to form on the plants. At the same time the leaves of the bottom of the stalks are cut away.

When the ears are formed some are picked for eating or for sale and the leaves around them are cut for use as fodder. Depending on location, some maize may be vulnerable to wind and, in some places, is doubled to protect ears from birds and the last rains, which otherwise rot the grain. The harvesting is in December.

The 1685 Kaqchikel and 1722 K'iche' calendars permit a more precise characterization of the annual agricultural cycle.

Table A3.1. Correspondence between the 1685 Kaqchikel and 1722 K'iche' calendars and the idealized agricultural cycle.

Kaqchikel 1685	Dates

1. Takaxepual
"Season of sowing the first milpas."
   "Name of an Indian month, the beginning of the year, or the time for sowing the first fields of maize" (Varea 1635; Annals of the Cakchiquels 1953:30; Coto 1983:522).
   Month fixed for the payment of tribute to the K'iche' (Annals of the Cakchiquels 1953:51).
	January 31, 1685– February 19

	February 20– March 11

3. Rukab' Tumuzuz
   Second tumuzuz (Annals of the Cakchiquels 1953:30).
   "Companion to the month of the flying ants."
	March 12– March 31

4. Qib'ixik "Season of smoke when one sows, or because there used to be the burning of brushwood before cowing, or as a metaphor, clouded or obscure because of the smog that there used to be." "Time of the year when the Indians sow" (Varea 1635; Annals of the Cakchiquels 1953:30).	April 1–April 20

5. Uchum
   "Season of re-sowing."
   "A good time for transplanting or planting vegetables" (Coto 1983); "sowing time" (Annals of the Cakchiquels 1953:30). All of the planted grains may not sprout and it is often necessary to replant. Replanting is usually done fifteen days after the first planting, although the amount of time between planting and replanting varies from a few days to three weeks, depending on the soil, climate, seed, pests, and so on. At times it may be necessary to replant several times when birds, animals, and insects are numerous or hungry.
	April 21– May 10

K'iche' 1722 A-II 1723	Dates	Idealized agricultural activity based on Bunzel (1952:51–53)
Che' Trees (Popol Vuh 1950:108)	January 19–February 7	February 1: Burning over of milpa and preparation of ground for new planting.
Tequexepual Time to plant the milpa (Popol Vuh 1950:108)	February 8–February 27	February 10: High winds; threshing of wheat. February 15: Early planting of corn. April 18: Rains begin.
Tz'ib'a Pop Painted mat (Popol Vuh 1950:108)	February 28–March 18	Rains.
Saq White like certain flowers (Popol Vuh 1950:108)	March 19–April 7	Rains.
Ch'ab' Muddy ground (Popol Vuh 1950:108)	April 8–April 27	April 10: Planting of potatoes. April 15: Cultivation of early white corn. April 18: Late planting of corn. Rains.
K'iso	April 28–May 2	May 1: Cultivation of early yellow corn. May 3: Commemoration of the Cross: "This is the day for giving thanks for the planting. It is at the season of planting. There are ceremonies in the hills for crops" (Bunzel 1952:418). Rains.

6. Nab'ey Mam "First season of the early-aged, because the milpa sown in this season does not grow high, and the creatures born in it as well." "Time of the premature old age, because corn planted at this time did not grow very tall, nor did infants born then." Both Mam months were considered unlucky: *Itzel k'ik ca uinak k'ih mam* (Evil days the forty days of mam), according to Varea (Annals of the Cakchiquels 1953:30). By this time the weeds in the milpa have reached such a size that a first cultivation or cleaning will be necessary. The plants are then about 20–25 cm tall and the weeds are overturned to a depth of 1–15 cm. With the abundant rains during this period of the year the maize plants and the weeds grow rapidly and a second cultivation will be needed. At this time the earth is hilled up about the plants to a height of 15 cm to protect them from the winds in late July and August. At this time ritual is often associated with maize cultivation. These rituals are selected by a daykeeper after consultation with the calendar. The number of cultivations is variable although in Chichicastenango the usual number is two. In the Cuchumatanes region to the northwest, it may vary from one to four, according to location and the type of seed (Stadleman 1940:113). In San Juan Chamelco the Q'eqchi' will usually weed four weeks after replanting. A second or third weeding will follow at four-week intervals (Wilson 1972:96–99). In ot	May 11–May 30
7. Rukab' Mam The second old man (Annals of the Cakchiquels 1953:30). "Companion of the season of the early-aged, for the same reason."	May 31–June 19
8. Likin Ka "Time when the earth is furrowed and slippery because of the many rains" (Annals of the Cakchiquels 1953:30).	June 20–July 9
9. Nab'ey Toqik "The first cut, wound, or bleeding; possibly an allusion to pruning or to incisions made in certain trees to extract sap. "First harvest of cacao or season of flaying." "First harvest" according to Crónica franciscana; "First harvest of cocoa" according to Ximénez (Annals of the Cakchiquels 1953:30).	July 10–July 29
10. Ruka' Toqik The second toqik (Annals of the Cakchiquels 1953:30). "Second harvest or companion of the former."	July 30–August 18
11. Nab'ey Pach First hatching, first incubation. "Time for setting broody hens." (Annals of the Cakchiquels 1953:30). "First season of hen-hatching."	August 19–September 7

Nab'e Mam First old man (Popol Vuh 1950:108)	May 3–May 22	May 10: Planting of wheat. May 15: Cultivation of early sweet corn; hilling of white corn. Rains.
Ukab' Mam Second old man (Popol Vuh 1950:108)	May 23– June 11	June 1: Hilling of early yellow corn. Rains.
Likinka Soft and slippery soil (Popol Vuh 1950:108)	June 12–July 1	June 15: Cultivation of late white corn. July 1: Cultivation of late yellow corn. Rains.
Ukab' Likinka Second month of soft and slippery soil (Popol Vuh 1950:108)	July 2–July 21	July 8: Hilling of early sweet corn; second hilling of early white corn. July 15: Hilling of late white corn; cultivation of late sweet corn; first potatoes available. Rains.
Nab'e Pach First time of hatching (Popol Vuh 1950:108)	July 22– August 10	July 25: Santiago: ceremonies in the hills for the World and for Santiago that he may not destroy the milpas with his horse (Bunzel 1952:419). August 1: Second hilling of early tallow corn; first hilling of late yellow corn. August 8: First hilling of late sweet corn. Rains.
Ukab' Pach Second time of hatching (Popol Vuh 1950:108)	August 11– August 30	August 20: Second hilling of late white corn; first ayotes and chilacayotes ripen. August 28: Second hilling of late yellow corn. Rains.

12. Rukan Pach
  Second hatching (Annals of the Cakchiquels 1953:30).
  "Second season of hen-hatching."
	September 8–September 27

13. Tz'ikin Q'ij
  Season of birds (Annals of the Cakchiquels 1953:30).
  Doubling consists of bending the stalks below the dried ear so that this will remain pointed downward, thus avoiding the entrance of water as well as minimizing the attacks of birds. According to Stadleman (1940:116) this is not generally practiced in the Cuchumatanes.
	September 28–October 17

14. Kaqan
  Hot season (Annals of the Cakchiquels 1953:30).
  "Season of red colors of yellow flowers."
	October 18–November 6

15. Ib'ota
  "Season of various red colors or of rolling up mats" (Annals of the Cakchiquels 1953:30).
  The end of the rainy season brings sprouting maize tassels, the augur of birds flying south, and heat around harvest time.
	November 7–November 26

16. Qatik
  Burning or clearing, or period of drought (Annals of the Cakchiquels 1953:30).
  "Passing or general sowing."
  Harvesting is usually done between October and December, depending upon the planting and the elevation. In colder regions it may extend into January. Bunzel (1952) states that harvesting in Chichicastenango is usually at the end of the November rains, and Tax (1953:46–52) gives December for Panajachel. Harvesting is done by husking; ears are stripped of their covering, with only one or two husks left attached to the shank. These serve to tie the ears in pairs for hanging. Another method of harvesting is snapping, when ears are merely torn from the stalk. According to Stadelman (1940:116), husking is common in colder regions and snapping in warmer areas of the Cuchumatanes.
	November 27–December 16

17. Iskal
  "Sprouting or to throw buds."
  Season when the Indians "sow in the mountain ridges," that is, in the highlands (Varea 1635; Annals of the Cakchiquels 1953:30).
	December 17–January 5

18. Pariche' "In the forest"; "time for covering or protecting oneself from the cold" (Annals of the Cakchiquels 1953:30); "season of blankets to protect oneself against the cold."	January 6–January 25
Tzapi Q'ij "Door that closes the year, day, and season."	January 26–January 30, 1686

Tz'isi Laqam The sprouts show (Popol Vuh 1950:108)	August 31– September 20	September 1: Second hilling of early sweet corn. September 8: Second hilling of late sweet corn. September 15: Beginning of harvest of early planting. September 20: Wheat harvest. Rains.
Tz'ikin Q'ij Season of birds (Popol Vuh 1950:108)	September 21–October 10	October 1: Harvest of small beans and potatoes. October 10: Doubling of stalks of late white corn. Rains.
Kaqam Red clouds (Popol Vuh 1950:108)	October 11–October 30	October 20: Doubling of stalks of late yellow corn. October 30: Doubling of stalks of late sweet corn. Rains.
B'otam Tangled mats (Popol Vuh 1950:108)	October 31–November 19	November 1: Harvest of ayotes and chilacayotes. November 11: San Martín: ceremonies in the house to give thanks for the harvest. November 15: End of rains; beginning of main harvest of late planting.
Nab'e Si'j First month of white flowers (Popol Vuh 1950:108)	November 20–December 9	
Ukab' Si'j Second month of white flowers (Popol Vuh 1950:108)	December 10–December 29	
Rox Si'j Third month of white flowers (Popol Vuh 1950:108)	December 30–January 18	January 10: Harvest of corn in cold regions; harvest of large beans in temperate zones.

NOTES

CHAPTER 1: INTRODUCTION

1. The K'iche'an languages, including K'iche', Kaqchikel, Tz'utujil, Sakapulteko, Sipakapense, Uspanteko, Q'eqchi', and Poqomchi', and Poqomam, are a well-defined subgroup within the Mayan language family and share phonological and grammatical innovations with other Mayan languages. The spatial distribution of K'iche'an languages encompasses the western and northern highlands of Guatemala, including parts or all of the departments of Alta and Baja Verapaz, Chimaltenango, El Quiché, Escuintla, Guatemala, Jalapa, Quezaltenango, Sacatepéquez, Sololá, Suchitepéquez, and Totonicapán. The languages of the calendars presented in this volume, K'iche' and Kaqchikel, together with Tz'utujil, share a close relationship, and together they comprise one branch of a subgroup of K'iche'an languages.

2. There is extensive correspondence in the University of Pennsylvania Museum Archives pertaining to a translation of the 1722 K'iche' calendar by Rudolf Schuller and Oliver La Farge (L. Satterthwaite; Hieroglyphic Research; Schuller–La Farge–Mason publication of Quiché mss.; Correspondence, University of Pennsylvania Museum Archives). The correspondence

begins in March 1929, when J. Alden Mason, then a curator at the University Museum, informed Samuel K. Lothrop, at the Museum of the American Indian / Heye Foundation, of the existence at the University Museum of the 1854 K'iche' manuscript by Vicente Hernández Spina. A few months later the museum director Horace H.F. Jayne showed the opening portion of the 1722 K'iche' calendar to LaFarge, who had just completed extensive ethnographic fieldwork at Santa Eulalia in the Cuchumatanes region of northwestern Guatemala. La Farge was able to translate the opening statement and quickly realized the historical significance of the manuscript. La Farge acquired a photographic copy of the 1722 K'iche' calendar and forwarded copies of the manuscript with a request for assistance with translation to several leading scholars, including such prominent Mayanists as Ralph L. Roys, Oliver G. Ricketson, and Karl Sapper. In June 1930 Mason visited Totonicapán in the Guatemala highlands and commissioned a local K'iche' speaker to translate the 1722 calendar. Mason then suggested to LaFarge that they consider the services of the Austrian philologist Rudolf Schuller, who was highly recommended by Franz Boas of Columbia University, Alfred M. Tozzer of Harvard University, and Alfred V. Kidder of the Carnegie Institution of Washington Maya Program. La Farge agreed and informed Jayne that he would arrange for a translation of the 1722 K'iche' calendar by Schuller. In January 1932 La Farge forwarded the completed translation by Schuller to Mason, who then sent it to Manuel J. Andrade at the University of Chicago and head of the Linguistics Program of the Carnegie Institution of Washington Maya Project. Andrade was unfamiliar with K'iche' but still sent Mason two pages of criticism of Schuller's phonological and orthographic analysis. Based on Andrade's critique, Mason reported that Schuller's translation was "even poorer than I thought." Schuller had to overcome a number of problems. Schuller had first transcribed the original text into his own phonetic system. In so doing he made "emendations," many of which seemed to be rather unjustifiable changes. His own notes state frankly that on the basis of comparative materials that he deemed conclusive, usually not K'iche' but sometimes Mexican or Nahuatl, he altered passages to make them conform. The excessive nature of these alterations was sufficiently apparent on first examination, even before it was determined that the work could not be published. The K'iche' used is somewhat archaic and clearly esoteric in nature. Some glosses have been lost from the language, others have changed the meaning, and many of the expressions, no doubt references to established religious ideas and esoteric practices, are extremely obscure. At the time no comprehensive vocabulary was available for translating K'iche' of this type into any European language. There was another problem, which Schuller could not help: although he spoke English tolerably well, his written English proved to be awkward, involved, and turgid. Thus, when it came to the ever-present difficulties of producing an accurate translation while retaining the idiom of the language into which the translation was made, he simply could not manage it. Some of his passages were unintelligible. Schuller died in 1932 and could not be called on to clarify or correct his own work, and neither La Farge nor Mason was competent to make a new translation or edit the existing one. In a letter to Mason, La Farge stated: "Schuller seems to have attacked the Maya problem, as a good many others have done, not as a scientific matter, but as a branch of the black arts. There is a kind of inbred, intense reality about his point of view which sometimes verges close to the borderline of insanity." In January 1933 La Farge sent revisions and added: "I also have here the abysmal translation which was made by that man in Guatemala [contracted by Mason in 1930]. It constitutes a large mass of waste paper as far as I am concerned. Do you want me to return it to you for the sake of completeness of your files, or shall I throw it away and save postage?" Upon careful review of the Schuller translation and the one commissioned in Guatemala, La Farge and Mason concluded that both were inadequate. They believed, however, that the original manuscript was sufficiently important to be published with a dependable rendering into a major language. La Farge then proposed to make a complete transcription of the original text in parallel columns with a copy of Schuller's version, so that readers could compare the two. LaFarge then undertook to teach himself K'iche' so that he could revise the Schuller translation. He also decided to do a much more ambitious commentary than had originally been planned and to prepare the entire manuscript for publication. The initial step, of teaching himself K'iche', proved to be impossible, and nothing resulted from the effort. However, a manuscript was eventually completed consisting of a transcription of the 1722 K'iche' calendar as recorded by Carl Hermann Berendt, a translation by Rudolph Schuller, and editorial notes and commentary by Oliver La Farge and J. Alden Mason. Although the manuscript was essentially complete and enhanced with printer's notations written in pencil throughout the text, the planned publication of the 1722 K'iche' calendar was never realized. Facing funding shortfalls caused by the Depression, La Farge abandoned the effort after it was learned that Ernesto Chinchilla Aguilar, a Guatemalan scholar, was preparing a Spanish-language translation. Unfortunately, Chinchilla Aguilar's translation was never completed. In 1959 the manuscript was deposited in the circulating collection of the library of the University of Pennsylvania Museum until it was transferred to the Rare Books section of the library where, together with the 1685 Kaqchikel and 1854 K'iche' calendars, it languished untouched for nearly fifty years.

3. A calendar wheel, possibly prepared by Berendt, is illustrated on folio iii of the 1722 K'iche' calendar. Three additional wheels are illustrated on folios iii–vi in the 1854 Hernández Spina calendar from Santa Catarina Ixtahuacan. The colonial dictionaries provide the following terms for calendar manuscripts: <vuh ahilabal k'ih> [wuj ajilab'al q'ij], book with which to count the days (Coto 1983:83); k'amuh [k'amuj], book that the Indians use to count the years, months, and days, as a calendar (Vico 1550).

CHAPTER 2: CALENDARIO DE LOS INDIOS DE GUATEMALA, 1685

1. Comment by Berendt states, "According to Torquemada, the second Mexican month is called Tlacaxipeualiztli."

2. Comment by Berendt states, "According to Torquemada, the eighteenth Mexican month is called Izcalli."

CHAPTER 3: CALENDARIO DE LOS INDIOS DE GUATEMALA, 1722

1. Marginal note, in later hand, reads, "por K. H. Berendt."

2. Marginal note, in later hand, reads, "Su idioma Sutujil segun Larraz y Juarros."

3. Marginal note, in later hand, reads, "Popol Vuh."

4. Marginal note, in later hand, reads, "conejo?"

5. The term *maceval* is a Nahuatl loan word from "macehualli," which means "subject, commoner, indigenous person" (Karttunen 1983:127). The term refers here to the general count of days as opposed to the count of the solar year.

6. The day and month names are given here in the original K'iche' form. The translation of day and month names leads to imprecision. The meanings of some terms are too opaque to find proper translations. In other cases, several layers of meaning are involved that are not reflected in a translation. Conjectured or possible translations and the renderings chosen in the previous translations of the text by Schuller (n.d.) and Edmonson (1965) are indicated in the annotations. The meaning of the day No'j is given as "idea, knowledge" (Ajpacaja Tum et al. 1996:226). Schuller translates it as "temper," and Edmonson (1965) gives "incense."

7. The meaning of the day Iq' is given as "wind, life" (Ajpacaja Tum et al. 1996:80). Schuller (n.d.) and Edmonson (1965) similarly translate it as "wind."

8. The meaning of the day Keej is given as "deer" as well as "four cardinal points" (Ajpacaja Tum et al. 1996:132). Schuller (n.d.) and Edmonson (1965) translate as "deer."

9. The meaning of the day E is given as "trace, path, destiny" (Ajpacaja Tum et al. 1996:75). Schuller (n.d.) translates as "tooth[-brush]." Edmonson (1965) likewise translates as "tooth." Schuller explains the translation by referring to the name of the twelfth day in the list of Metztitlán as *itlan*, "tooth," and in Central Mexico as *malinalli*, "a confused thing (*escobilla*)."

10. The meaning of the day Aq'ab'al is given as "dawn (*amanecer aurora*)" (Ajpacaja Tum et al. 1996:14). Schuller (n.d.) translates it as "house," and Edmonson (1965) more literally as "night."

11. The day K'at can be translated as "net, oppression (*red, opresión*)" (Ajpacaja Tum et al. 1996:157). Schuller (n.d.) gives it as "lizard," and Edmonson (1965) as "net."

12. Although Berendt's copy gives Cana throughout the text, the present-day reading of this day name is Kan. It can be translated as "serpent" (*serpiente emplumada y justicia*) (Ajpacaja Tum et al. 1996:121). Schuller (n.d.) as well as Edmonson (1965) translate as "serpent" or "snake." This day is said to represent justice.

13. The name of the day Kame is given as <Queme> throughout the manuscript. It can be translated as "death" (Ajpacaja Tum et al. 1996:124; see also Schuller n.d. and Edmonson 1965).

14. The verb *mixeqan* is an antipassive form that omits the object and stresses the agent. It is given in the immediate past and translates literally best as "the one who has just carried/borne/taken the burden."

15. The number "twenty" is used here to indicate the day on which the new month starts. In the Classic Maya calendar this is referred to either as the "seating of the new month" (e.g., *chum pop*, the seating of Pop) or as the "completion of the old month" (e.g., *k'al haab Yaxk'in*, "the completion of Yaxk'in). The internal structure of the K'iche' calendar does not leave any other option than the notation of the month, 20 q'ij Nab'e Mam, referring to the beginning of the new month and not to the last day of the old month. Most calendar-round notations in the text exhibit the grammatical pattern "[day] mixeqan chi 20 q'ij [month]." Edmonson (1965:118) translates the pattern as "[day] bears the 20 days of [month]," whereas Schuller (n.d.) renders it as "[day] and it already carried the 20 days of [month]." Neither translation takes the prepositional function of the marker *chi* into account. The construction *chi* + ordinal number + *q'ij* translates as "on the Xth day" (e.g., *chi rox q'ij*, "al tercero día") (Bocabulario en lengua 4uiche y castellana: fol. 18v). Accounting for the antipassive verb form the phrase may thus be translated literally as "the [day] who is the one who has just carried/borne/taken the burden on the twentieth day of the [month]," referring to the day as

the agent who shoulders the burden of time on the first day of the month. It seems plausible that in the K'iche' calendar the twentieth day indicates a "completion day," the day when the old month is completed and the new month begins. In modern K'iche', the number "twenty" is read as *ju-winaq*, "one human," when referring to days, or *ju-k'al* when referring to other countable objects. It is likely that in the calendar-round expression, the numeral likewise reads as *(u)-ju-winaq*, "twentieth," indicating completeness of time and human creation, possibly referring to the twenty fingers and toes of the human body.

16. The designation of the first month is translated by Schuller (n.d.) as "first old man" and by Edmonson (1965) as "first elder." The noun *mam* means "grandfather, elder, ancestor" but also "grandchild." It is also the K'iche' designation for the year bearer.

17. Q'anil translates literally as "yellowness" with various layers of meaning, including "ripeness." Schuller (n.d.) translates it as "rabbit"; Edmonson (1965), as "corn."

18. The root *toj* means "payment." It may be understood as "offering" (Ajpacaja Tum et al. 1996:402). Schuller (n.d.) translates it as "water"; Edmonson (1965), as "rain." Their translation is influenced by the fact that the day Toj in the K'iche' calendar corresponds to the Central Mexican day Atl, "water." Furthermore, the K'iche' patron deity Tojil is widely interpreted as being the rain deity (Edmonson 1965:124). Although the aspect of a sustenance provider connects this deity with rain, Tojil is primarily a lineage god associated with fire and sacrifice (cf. Akkeren 2000:176–181).

19. The day Tz'i', "dog" (Ajpacaja Tum et al. 1996:448; Edmonson 1965; Schuller n.d.), is generally associated with evil, negativity, and sexual impurity.

20. B'atz' can be translated as "thread" (Ajpacaja Tum et al. 1996:19). Schuller (n.d.) translates it as "ape," and Edmonson (1965) gives "monkey."

21. The meaning of the day name Aj is "cane, reed" (Ajpacaja Tum et al. 1996:1; Edmonson 1965; Schuller n.d.).

22. The day I'x may be interpreted as representing agriculture and environment (Ajpacaja Tum et al. 1996:83). Edmonson (1965) translates it as "jaguar"; Schuller (n.d.), as "sorcerer [jaguar]." Schuller adds: "Quetzalcoatl, the arch sorcerer, is very often accompanied by a jaguar. In the Pipil calendar of Guatemala the 14th day is called teyollocuani "the heart eater," i.e. the jaguar."

23. The noun *tz'ikin* means "bird" (Ajpacaja Tum et al. 1996:446; Edmonson 1965). Schuller (n.d.) varies his translation, defining it as "eagle."

24. The meaning of Ajmaq is not straightforward. Although modern *ajq'ijab'* generally interpret it as representing sin (Ajpacaja Tum et al. 1996:6), the root of the form is not *mak*, "sin," but *maq*. Schuller (n.d.) and Edmonson (1965) both translate it as "owl," based on the reading *ajmak* "sinner, owl" (Edmonson 1965:5).

25. Schuller (n.d.) translates Tijax as "flint-knife"; Edmonson (1965), simply as "flint." In modern usage the day is interpreted as representing suffering (Ajpacaja Tum et al. 1996:394).

26. Ajpacaja Tum and colleagues. (1996:130) give the meaning of the day Kawoq as "day of women and judges (*día de la mujer, los jueces*)." Schuller (n.d.) translates it as "rain"; Edmonson (1965), as "storm."

27. The literal translation of Junajpu is "one blowgun," which may designate the first hunter (Ajpacaja Tum et al. 1996:7; Edmonson 1965). Schuller (n.d.) translates it as "flower-rose," following its interpretation in the 1685 Kaqchikel calendar.

28. Added in different handwriting: *barrer la casa*, "to sweep the house."

29. The noun *imox* means "sword-fish" as translated by Schuller (n.d.). Edmonson (1965) translates it as "alligator." It is, however, unclear whether any of these are proper translations. Ajpacaja Tum et al. (1996:81) give the interpretation as "warning for stupidity (*advertencia para la idiotez y locura*)."

30. Literally, "second grandfather."

31. According to Edmonson (1965), the name of the month Likinka translates as "soft, slippery earth," referring to the time when the earth is soaked with rainwater. Although Schuller (n.d.) translates it as "soft of the hand," he notes that "soft soil" might be another adequate translation. The following month, Ukab' Likinka, reads as "the second Likinka."

32. Schuller (n.d.) reads *pach* as "incubation, or hen hatching," following the Spanish gloss in the 1685 Kaqchikel calendar. Edmonson (1965:84) follows their interpretation, rendering *nab'e pach* slightly differently as "first throwing." The root *pach* may also mean "to bend over," indicating an act of humility or penitence. Read as *pach'*, "mashing, crushing," the name of the month could relate to some other form of agricultural activity. The following month, Ukab' Pach, reads as "the second Pach."

33. Coto (1983) gives *tz'is* as "top of a pole" and *laqam* as "flag (*bandera, estandarte*)," which is probably best translated as "hoisting flags." Schuller (n.d.) translates Tz'isi Laqam as "sewing flags" and associate the meaning of this month with the month Panquetzalitztli of the Mexican calendar. Edmonson (1965:135) translates *tzisil lakam* as "shoots come" and thus offers an agricultural interpretation.

34. The verb *kub'axik* signifies "to take office, to assign somebody to an office" (Ajpacaja Tum et al. 1996:146–147).

35. Tz'ikin Q'ij literally means "bird day," which has been translated more generally as "time of birds" by Schuller (n.d.) and "bird season" by Edmonson (1965).

36. The month Kaqam is translated by Schuller (n.d.) as "time of red colors and of yellow flowers," following the Spanish gloss in the 1685 Kaqchikel calendar. Edmonson (1965:56) renders it as "red clouds."

37. Coto (1983) gives *b'ot* as "to earth up plants (*aporcar las plantas, arrollar*)." The month B'otam may thus be the month of "earthed-up plants." Schuller (n.d.) and Edmonson (1965) translate it as "rolling up mats."

38. The noun *si'j* is a variant form of *kotz'i'j*, "flower"; Coto (1983) gives *sih* as "tree, flower." Schuller (n.d.) translates it accordingly as "first flower" and the following months Ukab' Si'j and Rox Si'j as "second flower" and "third flower."

39. The noun *che'* means literally "tree, wood"; Schuller (n.d.) and Edmonson (1965) translate accordingly.

40. The name for the month Tekexepoal is a loan from the Nahuatl word Tlacaxipe-hualitzli, which means "the flaying of men"; Schuller (n.d.) and Edmonson (1965) translate accordingly.

41. The name of the month Tz'ib'a Pop translates literally as "painted mat"; Schuller (n.d.) and Edmonson (1965) translate accordingly.

42. The adjective *saq* means "white"; the meaning of the month has been translated accordingly by Schuller (n.d.) and Edmonson (1965).

43. The noun *tab'al* derives from the verb *ta'o*, "to hear, to ask," and literally translates as "instrument/place of hearing/asking." It is used to refer to an offering, an altar, or any sacred place where the *chuchqajawab'* speak their petitions and prayers (León Chic 1999:228).

44. The noun *ch'ab'* means "arrow, bow"; Schuller (n.d.) translates accordingly.

45. Literally, "its ending." It is unclear if this phrase is a variant way of indicating the five remaining days that are otherwise referred to as *tz'api q'ij*, "closing days," or if it should be read literally as referring to the year's ending.

46. The noun *tz'api q'ij* is composed of the positional root *tz'ap*, "closed, shut," and the noun *q'ij*, "day." Garrett Cook (personal communication) notes that the formula "closing day" may refer to "the physical closing of the altars of the family groups and sodalities" as they take place in Momostenango.

47. The original text has *mixhue*, which is a copying mistake by Berendt. As the following lines have the verb *mixuk'a'n* (*mixu4an* or <mixu4am> in the Berendt copy), this verb form has been chosen here for correction. The antipassive verb form translates literally as "the one who has sustained (with hands)/held up."

48. The given date is incorrect and should be May 3 instead.

49. Marginal note, in later hand, reads "11 ag!"

50. Marginal note, in later hand, reads "31 ag!"

51. Marginal note, in later hand, reads "20 Sept!"

52. Marginal note, in later hand, reads "10 oct!"

53. Marginal note, in later hand, reads "30 oct!"

54. Marginal note, in later hand, reads "19 mar!"

55. Marginal note, in later hand, reads "8 abr!"

56. The classifier *ichal* always combines with a numeral in first position. Ximénez (1985) described the meaning of the nominal form as "particula numeral para colectividades" (numeral particle for groups) and gives <cab-ichal> as "nosotros dos." Dürr (1987:117) gives *ichal* as "group of X numbers" (i.e., *ukab' ichal*, "his group of two"; *ki-kaj ichal*, "their group of four"; and so on). Edmonson (1965:45) renders *kaj ichal* as "by fours." *Kab'-ichal* is best translated into English as "both." *Ro' ichal* as a term addresses a collection of five.

57. *Kaq-wach* literally means "red-face." Ximénez (1985) gives <ɛaɛwachih> as "odiar, envidiar."

58. *Mox-wach* literally means "left-face." Here it forms a semantic couplet with *kaqwach*, "red-face." Ximénez (1985) gives *moxvach* as "envidia, envy."

59. The term *ka'ib' u-k'ux*, "two (is) his-heart," is regularly used in modern K'iche' to indicate that someone is undecided. Edmonson (1965:126) translates the term as "those of the crooked hearts." In modern K'iche' prognostication, the metaphor "of two hearts" is used to refer to a state of emotional instability, hypocrisy, and desperation (León Chic 1999:223).

60. The reconstruction follows Schuller's (n.d.) and is consistent with comparable contexts within the text. Edmonson (1965:126) reads *sina'j una'oj* as "scorpion is their advice."

61. The couplet *u-k'ux u-na'oj*, "his heart, his mind," occurs several times in the prognostications of this first divinatory calendar. It refers to the condition and meaning of the day.

62. Schuller (n.d.) translates this phrase as "he put the little mantle upon his neck." The translation is based on vocabulary entries in Basseta (1698) giving <uchai ukul> as "collar (*ruedo del manto*)," and Morán (1720) (Poqomam) giving <chay> as "thin blankets that are worn (*unas mantillas degadas que se visten*)."

63. Coto (1983) gives *k'oyim* as "tangle (*marañar*)," which Schuller (n.d.) rendered as "loitering about." Ximénez (1985) has *coyin* as "kick, punch (*dar con pie o puño*)," analyzed

by Edmonson (1965) as *q'oyij*, "to beat, strike, blow." The second complement of the couplet is equally opaque. The noun *sis* means "pizote" as much as "chicken louse," but Coto (1983) also translates it as "anxious person (*inquieta persona*)." Schuller translates it as "gossiping," which Edmonson (1965) gives as *zizipar*.

64. Either Berendt has mistakenly copied the phrase twice or he neglected to complete the first phrase, which seems to be missing the negative particle *ta* and possibly a nominal form parallel to *ub'anom* in the second phrase.

65. Literally, "daughter-in-law-ness" and "nobleman-ness." The latter term probably means "son-in-law" or "groom." Coto (1983) gives <aqanimak>, "casarse, girifalte, halcón, hidalgo, sacre"; the Calepino de Cakchiquel (Sáenz de Santa María 1940) has an entry for <aqanimaki>, "noblemen, authorities (*los principales, autoridades*)."

66. *Q'ali-b'al* literally means "place/instrument of revelation of appearance." In the present text, the noun generally occurs with the preposition *chuwach* and is therefore translated as "throne." Schuller (n.d.) translates it as "under the protection," following Ximénez (1985: f. 36v), who gives *oj k'o wi xe' a-muj, oj k'o wi xe' a-q'alib'al* (we are under your canopy, we are under your throne) as "we are under thy protection." Furthermore, Schuller (n.d.) points out that *q'alib'al koj, q'alib'al b'alam*, "the throne of the puma, the throne of the jaguar," was an expression referring to the principales or kings.

67. It is likely that this is an agentive form as the first part of the couplet. The Calepino de Cakchiquel (Sáenz de Santa María 1940:305) gives the noun <patal> as "tribute, load of salt (*tributo, carga de sal*)." *Patan*, in general, is "service, office, tribute."

68. The phrase *saq amaq'* (white nation) is a metaphor for peace (Edmonson 1965:158). Coto (1983) defines it as "state of glory (*estado de gloria*)." The agentive prefix *aj-* translates literally as "he of" but refers to the "maker of something." Schuller (n.d.) translates the form as "peace-makers."

69. The form is derived from the verb *ta'nik*, which means "inquiry, action of asking" as well as "to listen." It literally means "instrument/place of inquiring/listening." Schuller (n.d.) translates it as "hearing."

70. The form *u-k'axtok'-a-b'al* derives from *k'axtok'*, "demon (*malvado, demonio*)," or, according to Ximénez (1985), "deceiver (*engañador*)" and translates literally as "place/instrument of the demon." Edmonson (1965:98) points out that *k'axtok'* was the demon who according to Kaqchikel myth caused the dividing of the tribes. Basseta (1698) gives *caštocobal*, "lie, falsehood"; Coto (1983) translates *k'axtok'ob'al* as "lie (*mentir*)." Schuller (n.d.) chooses the term "deceiving."

71. The root of this antipassive verb is *ch'ak*, "flesh;" the transitivized form *ch'akuj* translates as "to eat flesh/meat."

72. The translation of this phrase is ambiguous. The form <boh> may also be read as *b'oj*, "clay pot," which would render the translation as "tight is the top of the pot." Schuller translated it as "the cover right side up in the pot." The form *uwi'*, however, is not preceded by the standard preposition *chi* or *pa*, as expected in this context. Thus, *boh* should be read as *b'oj*, "cotton fabric (*tela de cotón*)" (cf. Ajpacaja Tum et al. 1996:33). The semantic context may be related to the tightening of the *su't*, the head scarf used by the priest-shamans, or even to the headbands used by Classic Maya rulers. The reading as *b'oj* also seems to bear more sense with the following verb *kolij*, which signifies "to tear down the ribbon, or batten (*quitar cinta o listón de la cabeza*)" (Ajpacaja Tum et al. 1996:138). Within the general prog-

nostication of the day, the "taking down of the head scarf" might refer to the ending of an office or even the death of a leader, similar to the metaphor from Classic Maya texts.

73. The form *4aceinnak* seems to make most sense if read as *k'asun-inaq*, which derives from *k'asuj*, "to awaken somebody."

74. Marginal note: "julio 2/abril"

75. Literally, "feather-ear."

76. Original text uses the traditional abbreviation *xptis* for "Christ."

77. The translation of this passage is not definite. The verb may derive as an antipassive from the noun *sok*, "nest, bed of straw (*nido*)." It is grammatically unusual that the verb ends in *-ik*, which should only occur at the end of a phrase. The comparative prognostic in the second divinatory calendar has the form *sukun*. The verb, however, might also derive from *sokik*, "to hurt," which would also bear sense in the context of the entire phrase given that "serpent" may function as a metaphor for "disease."

78. Coto (1983) gives <cak beleh> as "serpent (*culebra*)." The Calepino de Cakchiquel defines *qakabelom* as "wound, sore" (Saénz de Santa María 1940:181). Taking into account that in modern K'iche' "serpent" is used as a metaphor for rheumatic diseases, a comparable semantic context is rather likely.

79. The repetition of the particle *wi* is likely a copying mistake.

80. The translation of this form is not definite. Basseta (1698) gives *taz* as "orden" and "dividir"; he translates the phrase *xnutas waqan* (literally, "I divided my leg") as "I betrayed (my husband)." The connotation is clearly sexual. Schuller (n.d.) thus translates as "adulterous."

81. The translation is literal, although it is not entirely clear whether the text actually refers to the creator god Juraqan or the couplet "divider—one leg" is meant to indicate sexual sin as a prognostication.

82. This passage in the text states the purpose of the divinatory calendar as a means for prognostication and instruction about when to hold a ceremony and present offerings. Schuller (n.d.) translates the phrase less literally as "and that day in the day-reader [will tell us] . . ."

83. Here Berendt has written "de otra mano." The different style of the following divinatory calendar confirms that this text is of different regional origin and has been written by a different author.

84. The manuscript text has <vtisah>, which is clearly a copying error.

85. The first letter is not clear; the reconstruction as *tze'n* follows Schuller (n.d.).

86. The last letter is not clear; *utz*, "good," seems to fit the semantic context.

87. The reading of *ɛoh* as *q'ojom* and the translation are not definite. Schuller (n.d.) suggests a copying mistake and reads *xa kuchul chi q'oj*, which he translates as "covered with the mask." Coto (1983) gives *q'oj* as "hunt, lazo (*cazar animales, lazo*)."

88. The author, or scribe of the second divinatory calendar, uses *us* in several places to indicate *utz*, "good." Thus, Schuller (n.d.) translated it as "first it was all right." The couplet structure and the indication of a long vowel in *xaan* (i.e., *xa'n* or *xaan*, "mosquito") suggests the above reading, the semantic context of which, however, is opaque.

89. Coto (1983) gives *ch'utin nu qux* as "thin, miserable (*flaco y miserable*)" and *ch'utin ru qux* as "fragile person (*delesnable persona*)." The Calepino de Cakchiquel (Sáenz de Santa María 1940) gives the form as "disheartened (*pusilánime*)," which is why Schuller (n.d.) translates it as "coward."

90. The original form in the text is *itzilah*. The context suggests that this reads *utzilaj*, "very good," and not *itzelaj*, "very evil." The spelling in the following paragraph has been adjusted accordingly.

91. Added to the day 5 Q'anil is the Gregorian date "13 de Marzo de 1770 @."

92. This phrase literally translates as "the strength, the manliness of the day."

93. The Calepino de Cakchiquel (Sáenz de Santa María 1940) gives <qolobal> as "freno, brida, atadura," and Basseta (1698) has *colob* as "cordel, soga." The suffix *-al* indicates an abstractive noun. The forms are unusual but seem to be valid in the given context.

94. Berendt copied *cic*, which may have been *çic* in the original manuscript and could either be read as *sik'*, "call, cry," or *xik*, "hawk," which appears to be more likely in the given semantic context.

95. Schuller translates *chiya'* as "he shall place it." Both translations are valid.

96. The metaphor *el uk'u'x* is given by Ximénez as "sympathetic" (as quoted in Edmonson 1965:33).

97. Coto (1983) gives the phrase *ixoq chi'j* as "to be blasphemous (*blasfemar*)"; it literally translates as "woman speaking." The form *ixoqichinel* is an abstractive noun with an antipassive root, which may be best translated as "blasphemy."

98. The manuscript gives *eko*, which has been corrected based on the comparative prognostic in the first divinatory calendar, which gives the form *eqa'y*, "bearer."

99. *To'l* literally means "helper, defender." Edmonson (1965) describes the *to'l* as a small red bird that whistles on the roads.

100. In the given context, *ral* likely refers to another type of bird. It may also function as an attribute to the *tapi cholol* (e.g., *ral tapi cholol*, "the child of tapi cholol"). Ximénez (1985) gives *ral* as "to hunt for birds (*cazar pájaros con redes o lazos*)."

101. Edmonson (1965) identifies the *tapi chol* as the "crab sheller bird."

102. Coto (1983) defines *xco* as "parrot (*el papagayo*)."

103. The form most likely derives from *q'e'tz*, "viejo, personas." Basseta (1698) gives *quetzam*, "difunto"; *quez*, "cosa vieja y aspera"; and *quezarik*, "hacerse agudo."

104. The metaphor *xikin uwach* literally translates as "the ear is its aspect." Schuller (n.d.) translates the phrase as "attentive."

105. Berendt copied *nisma ah*. The reconstruction as *nimalaj* is based on the semantic context. The noun *loq'ob'al* means literally "instrument/place of loving/buying." Basseta (1698) gives for *loɛobal* the meaning "benefit (*beneficio*)"; Coto (1983) translates it as "mercy (*merced*)."

106. The verb *mayik* means "to think, ponder." Coto (1983) also gives *maih* as "admirable thing, authority, benign." However, *may* may also translate as "tobacco"; and Ximénez (1985) gives *may* as "twenty years." The verb may furthermore signify "to weed," possibly yielding "the weeding of days" in the given context.

107. This form can be translated literally as "the fun-shaken." The root *pay* means "joke, fun." The second part of the compound is given as *tatom* in the manuscript text, but it is most likely read as *totom*, which derives from *toto'j*, "to shake." The translation as "buffoon" follows Schuller (n.d.).

108. Coto (1983) gives *tze'b'al chi tzij* as "word-play (*juego de palabras*)"; Edmonson (1965) has *tze'lej tzij* as "joke" and *tze'le'l tzij* as "joker."

109. Ximénez (1985) gives the phrase *k'is uk'ux* as a metaphor for "to be bored."

110. Schuller (n.d.) suggests a copying mistake by Berendt and reads *ajb'is* as "the master/he of sorrow."

111. Repetition of *chi* in the original manuscript.

112. The noun *jom* is given by Edmonson (1965) as "ballcourt"; Ximénez has it as "graveyard, ballcourt (*cementerio, patio donde jugaban al batey*)."

113. The translation of this form is not quite grammatical. Given the context of the prognostic in the first divinatory calendar, the translation as "listening" seems to be the closest. Coto (1983) gives the phrase *qhalaqhak ru xiquin* as "oreja"; Edmonson (1965:24), citing Brasseur de Bourbourg, gives *chalab'aj*, "listen." Grammatically, the ending *-C₁-oj* derives an adjective form with moderative connotation, that is "half listened." The basic meaning of the verb *ch'al* is "to flay," which however does not seem to fit the semantic context of this phrase. Schuller (n.d.) translates the entire sentence as "the owls will attentively listen at the head of their way."

114. Coto (1983) gives *puvakixic* as "to gild something (*dorar algo*)." Since *puwaq* properly means "silver" and not "gold," I have chosen the translation of "silver-making." Schuller (n.d.) translated it as "making jewels."

115. The form *ik'o q'ij* translates literally as "passes before sun" and refers to Venus as a morning star.

116. Coto (1983) gives *elah chijn* as "surrender (*rendirse*)." The translation assumes that the phrase *chelech chin* in the manuscript text is a mistake by the copyist.

117. The literal translation of this passage is "his wish already is his life."

118. This translation follows the translation of the prognostic for the same group of days in the first divinatory calendar. Schuller (n.d.) translates *sukun* (*zocon*) as "is extended."

119. Schuller (n.d.) translates the concept metaphorically as "discord."

120. Schuller (n.d.) translates *rax* here as "humid," which bears sense as it occurs in a couplet with *chaq'ij*, "dry." Here I have chosen the more literal translation.

121. It is not clear how this should be translated.

122. The verb form *chuchu-x-inaq* derives from the noun *chuch*, "mother." The verb *chuchuj* means "to take as mother (*tomar madre*)."

123. The function and meaning of the second ending *-aq* in *qajawinaq-aq* is unclear. As this term forms a couplet together with the preceding noun *chuchuxinaq*, the context confirms the grammatical analysis as a participle derived from the transitive verb *qajawij*, which Ximénez (1985) gives as "take as father (*tomar padre*)."

124. A small circle with two pairs of perpendicular intersecting lines follows "11 ymos."

CHAPTER 4: CALENDARIO DE VICENTE HERNÁNDEZ SPINA, 1845

1. Hernández Spina translates the term *aj q'ij* erroneously as "Sacerdote del Sol"; the K'iche' word *q'ij* means "day" as much as "sun."

2. Brasseur de Bourbourg (1871:83) writes: "Hernández Spina (presb. D. Vicente). Kalendario conservado hasta el dia por los sacerdotes del Sol en Ixtlavacan, pueblo desceniente de la nacion K'iche', descubierto por el Prebitero Vicente Hernández Spina. Santa Catarina Ixtlavacan, agosto 12 de 1854. gr. In-fol. Obl, plié. Manuscrit en 13 fol., d'une main fort soignée. Le calendrier est identique avec l'ancien système quiche. Ce que ce document

offer de curieux, ce sont les noms des lieux où les Indiens de Santa Catarina Ixtlahuacan vont sacrifier, ceux de leurs prêtres, ainsi que priers qu'ils adressent au Soleil, mélange de souvenirs de leur idolâtrie et d'idées chrétiennes."

3. Direct translation of the K'iche' text.

4. Literally, "he gives it."

5. Ik'o Q'ij, "Passes before Sun," referring to Venus as the morning star.

6. *Tz'ikin,* "bird."

7. *No'j,* "idea, reason."

8. The 1:50,000 scale topographic maps for Santa Catarina Ixtahaucan and Nahuala are dense with place-names. The real density is even greater. Within his or her radius of frequent travel, a K'iche' can specify locations to within a few tens or hundreds of meters of trail by naming an aldea, caserío, finca, or hacienda and the name of a spring, cliff, cave, or bridge. Most place-names in K'iche' and Kaqchikel are based on plant, animal, and mineral names prefixed with the modern locative affixes (*chi,* at, in, on; *chuwa,* in front, opposite; *chuwi,* on, upon, over, above; *pa,* at, in, on; *xe,* under, beneath). Nearly all place-names are translatable, although the hispanicized spelling given on topographic maps does not always allow translation. For example, Kyaq Pa Tzulub', "red (blood) at the pole"; Edmonson (1965:136) translates this as "arrow dance" and indicates that it is a place near Lake Atitlán. Quiacbab (Quiacbatzulub, Quiacuetzulub) is a hill (cerro) between Nahuala and Santa Catarina Ixtahuacan, Solola; locally known as Quiacuetzulub, also known as Paculam.

9. Retal Ulew, "Limit of the Land"; a forest that borders the southwest boundary with Santo Tomás.

10. San Gregorio, "Otro bosque que se ha formado en donde existio el pueblo de este nombre? E contiguo a San Miguelito" (Hernández Spina 1854).

11. Chi Masat, "Place of the Deer"; Nahuatl *mazatl,* "deer." Chirijmasa is a river boundary between Nahuala and Zunil as well as a hill (cerro), Santa Catarina Ixtlahuacán.

12. Santo Tomás (Pecul) is a volcano near Chuisantotomas in the caserío of Guineales, Santa Catarina Ixtlahuacan.

13. San Antonio, a mountain near Chuisanantonio, Nahuala.

14. Chuwa(ch) Tulul, "opposite of the zapote tree." Chua-tulul is a mountain near Santo Tomás Suchitepequez.

15. Tza'ma Yak, "end of(?)"; Edmonson (1965:152) identifies Tama Yak as a town mentioned in the indigenous sources. Samayac is a Kaqchikel-speaking municipio in the department of Suchitepequez, located south of Santa Catarina Ixtlahuacán.

16. Chuwi' Poj, "on top of the hill"; in the municipality of Santa Catarina Ixtahuacan.

17. Pa Ra(x) Che', "place of the green tree," Parrache (Parrachac[?]), paraje, Santa Catarina Ixtahuacan.

18. Chuwi' Pek (J)ul, "above cave hole"; note that "Pekul" is also the name of the Acatenango volcano. Chui-pecul, Pecul, or Santo Tomás is a volcano in the departments of Quezaltenango and Solola; Pecul is a hill near Santa Catarina Ixtahuacan.

19. Xe' Xak, "below the cliff"; place near San Cristóbal Totonicapán (Edmonson 1965:146). Xexac is a mountain slope in the municipal jurisdiction of Santa Catarina Ixtlahuacán and a caserío in the aldea Xejuyub.

20. Chuwa(ch) Jolom, "opposite of the head"; Chua jolom, un lugar que esta en la mediania del volcan del Zunil, en donde existe una cabeza de figura humana de piedra.

21. Pa Ch'ib', "place of the witch-doll tree"; Pachip, Pachipac, caserio, aldea Chirijox, municipio Santa Catarina Ixtahuacan; caserio municipio Nahuala.

22. Pa Saqi Juyub', "place of the white mountain"; Pasaqui-juyup, Pasaquijuyub, caserio, aldea Tzanjuyub, Santa Catarina Ixtahuacan.

23. Chuwi' Q'ax Tum, "above the crossing(?)"; Chui-caxtum, possibly, caserio in aldea Chiyax, municipio Totonicapan.

24. Chuwi' Ixkanul, "on top of the volcano"; Chui-ixcanul, ancient name for Volcan Santa Maria near Quezaltenango.

25. Chuwi' Lajuj Juyub', "on top of ten mountains."

26. Ix Kab' Yut, "lady of the honey bundle."

27. Chuwi' Saq K'ichi', "above the white forest."

28. Chuwi' Nima Juyub', "on top of the great mountain"; Chui-nimajuyup, Chuani-majuyup, montaña, Nahuala, Solola, near Santa Lucia Utatlan, 3,092 m ASL.

29. Pa Saj Kajá, "place of the thunder hillside."

30. Xe' Tzalam Choj, "below the echo-slope."

31. Chuwi' Ukub' Krus, "above the seat of the cross"; Chui-ucup-cruz, a forest in the Siete Cruces mountains west of Santa Catarina Ixtahuacan.

32. Chuwi' Muru K'im, "above the curly grass."

33. Sachb'al, "place of doubt."

34. Pa Rax K'im, "place of the green grass"; Parraxquim is a caserio in the municipio of Nahuala, Solola.

35. Chuwi' Tun Ab'aj, "above the drum stone"; Chui-tunabaj, mountain in the municipio of Totonicapan; and possibly Tunaquel, mountain in the municipal jurisdiction of Nahuala; it provides the drinking water for Nahuala.

36. Chuwi' Patan, "above Patan"; Chui-patan is a hill in the municipal jurisdiction of Nahuala, near Santa Catarina Ixtahaucan.

37. Chu Ja' Kiej, "at the deer river."

38. Chuwi' Pe Akul, "above(?)."

39. Xe' Pe Akul, "below(?)"; Xe-peacul is a place in the aldea Xepiachul, Santa Catarina Ixtahuacan; Piacul is a mountain near Santa Catarina Ixtahuacan.

40. Chuwi' B'ital Amaq', "above the formed nation."

41. Chuwi' Saq Ulew, "above the white land."

42. Xe' Ka K'ix Kam, "below the thorn vines"; Xecaquixcam (Xecaquixcan) is a caserio of aldea Xepiacul, Santa Catarina Ixtahuacan.

43. Chuwi' Q'umil, "above the squash field."

44. Pa Kox Koj, "where the puma strikes."

45. Chuwi' Saq Ab'aj, "above the white stone"; Chui-sac-abaj or Chuisacabaj is a caserio in the aldea Tzanpuz or Tzanpoj (Tzanjuyub), Santa Catarina Ixtahuacan; Sacabaj is a mountain in Santa Catarina Ixtahuacan.

46. Ch'uti K'im, "place of the small grass."

47. Chi Q'alib'al, "at the place of revelation of appearance"; Xecalibal is a caserio in the aldea Tzanjuyup, Santa Catarina Ixtahuacan.

48. Chuwi' Sibil, "above the smoke"; Chui-sibel is a caserio in the municipio of Santa Catarina Ixtahuacan; Chuasubel is a mountain near Santa Catarina Ixtahuacan.

49. Chuwi' Kab'om, "above(?)"; q'abun, "plowed land."

50. Wal Kox, etymology unclear; Ual-cox, Río Ugualcox, headwaters are north of montaña Siete Cruces in the sierra Parraxquim; it flows into the Río Nahualate north of caserio Pala.

51. Chuwi' Cholochik Chaj, "above the aligned pine trees."

52. Tz'am Ab'aj, "edge of the stone"; Tzam-abaj or Tzamabaj is a mountain within the municipal jurisdiction of Santa Catarina Ixtahuacan.

53. Pa Nawal (J)a', "place of spirit water"; Naguala, K'iche'-speaking municipio in department of Solola, twelve kilometers from Santa Catarina Ixtahuacan.

54. Pa La', "place of the Chichicaste"; Pala, caserio, aldea Guineales, Santa Catarina Ixtahuacan; aldea, Santa Catarina Ixtahuacan.

55. Tza'm Pojon, "edge of the hill(?)"; according to Edmonson (1965:133), a place near Quetzaltenango; Tzampo or Tzanpoj is an aldea within the municipio of Santa Catarina Ixtahuacan.

56. Pa Q'aman Uj, "place of the marker(?)."

57. Pa Tzima Ju(t)i(n), "place of the gourd tree"; according to Edmonson (1965), a place near Santa Catarina Ixtlahuacán.

58. Pa Xe' Tza'm, "under the edge."

59. Pa Maxan, "place of the large leaf"; according to Edmonson (1965:71), a place near Cubulco and Santa Lucía Utatlán.

60. Pa Raxa B'on, "place of the green(?)."

61. Chuwi' K'is Talin, "above(?) Catarina"; *quistalín* is a Spanish loan; "cristalín" and the meaning of the place-name is "above clear (water)."

62. Chuwi' Kyaq K'ix, "above the red thorn."

63. Chuwi' Pa Sib'il, "above the smoke."

64. Sab'al Tuney, "(place of) drying dahlia flowers"; Sabal-tuney, municipal boundary marker, Santa Catarina Ixtahuacan.

65. Chuwi' Tz'ukub'al, "above the seat"; Chui-sucubal, Tzucubal, aldea, Santa Catarina Ixtahuacan; Tziquin Sucubal, municipal boundary marker.

66. Etymology unclear; Pa-Yocaj, possibly Payoxaja, caserio de la aldea Xejuyub, municipio Nahuala, Solola.

67. Chuwi' Q'enom Kajul(ew), "above wealthy sky-earth."

68. Chuwi' Jox, "above the scraped pine tree"; Chui-jox, Jox, mountain, 2,985 m ASL, in the municipal jurisdiction of Santa Catarina Ixtahuacan; Chuajax, municipal boundary marker.

69. Chuwi' Q'oxom, "above the lookout"; Chui-coxom, Chuicoron(?), montaña, Santa Catarina Ixtahuacan; Xecoron, montaña, Santa Catarina Ixtahuacan.

70. Pa Q'an Ab'aj, "place of the yellow stone."

71. Chuwi' Pa Ch'ek, "above the knee."

72. Q'ajinaq Krus, "fallen cross."

73. Chui-Maria-Tecum, Maria Tecun, paraje, municipal jurisdiction of Totonicapan.

74. Chuwi' Muchulik B'aq, "above the crumbled bones."

75. Chuwi' Pop Ab'aj, "above the mat stone"; Chua-pop-abaj, Chuapoj, cerro, Nahuala, near Santa Clara La Laguna and Santa Catarina Ixtahuacan.

76. Chuwa(ch) Ja' Pa Xot, "opposite of the well/river at the tile."

77. Chuwi' Ukub' K'ajol, "above the seat of the son."

78. Pan Q'ix, "place of thorns."

79. Chuwa(ch) Mansa'n, "opposite of the mansana (terrain)"; Chua-mansan (Chua-manzan), caserio in the municipio of Zunil.

80. Chuwi' Peraj Xajab', "above the sandal-shoe."

81. Chuwi' Tziki Che', "above the willow tree."

82. Chi Jul, "place of the hole."

83. Chuwi' Kyaqa Siwan, "above the red ravine"; Chui-quiacsiguan, Quiacsiguan, caserio, Nahuala.

84. Chi Cha K'im, "place of the bitter grass."

85. Pa Yajut, "place of(?)"; place near Nahuala; Payajut, mountain in the municipal jurisdiction of Nahuala near the municipio of Santa Lucia Utatlan; paraje in the aldea of Chucchexic, Santa Lucia Utatlan.

86. Xe' Pur, "below the donkey/teeth"; the translation is not entirely clear but could refer to a place near Totonicapán where there is a stone that looks like a donkey.

87. Chuwi' Sempoal, "above Sempoal"; Nahuatl *cempohualli*, "twenty."

88. Translation of the Spanish text.

88. Pa Aj, "place of the cane."

89. Chui-Santa Lucia, Santa Lucia Utatlan(?), municipio adjacent to Nahuala and Santa Catarina Ixtahuacan.

90. Pa K'ix Kab', "place of the thorn hut."

91. Literally, "white sons and daughters."

92. Literally, "claws."

93. Repeated in the text, ma ta.

94. Translation of the Spanish text.

95. Sija Raxquim, Sija, "es la altura mas culminante que hay en los terrenos de Santa Catarina, hacia al norte de dicho pueblo."

96. Chuwi' K'ixik, "above the thorns."

97. Chiri' Pe Akul, "place of the(?)."

98. Pa Kaja, "at the hillside."

99. Raqan Taq'aj, "along the valley"; according to Edmonson (1965:117), a place near San Andrés–Xecul and Totonicapán.

100. Pa Ch'ipaq, "place of the soap-root"; a caserio of aldea Chirijox, Santa Catarina Ixtahuacan.

101. Xe' Pa Tuj, "below the place of the sweat-bath."

102. Pa Ximb'al, "place of tying."

103. Chuwi' Kuwil, "above the covering."

104. Chi Raxon, "place of the blue dove."

105. Xe' Chojoj Che', "below the rustling tree."

106. Siete Cruces, mountain, Nahuala, Santa Catarina Ixtahuacan and Cantel, 3,300 m.

107. Tz'ib'a Che', "painted tree."

APPENDIX I: NOTES ON HIGHLAND MAYA
DIVINATORY CALENDARS, BY ROBERT BURKITT

1. Some of the information contained in Burkitt's notes is provided in code. Unfortunately, the key to the code is unknown. As far as can be discerned the numbers included

in parentheses seem to refer to pages in his notes and the source of his information. This information has been deleted from the text. In a field notebook titled "Comparativ [sic] Table," Burkitt lists approximately 2,800 terms with translations, including day names (ff. 21–23) and month names (ff. 23–24). Unfortunately, there is no date of compilation, although Burkitt states that he had extracted some information from a book that had belonged to Gustav Kanter, the administrator of a large property in the municipality of Nenton, northwestern Guatemala. Kanter maintained good relations with local and national politicians until he began to have commercial negotiations with Mexican revolutionaries along the border between Mexico and Guatemala. He was accused of selling firearms and harboring revolutionaries at his finca house at Chacula. In 1915 an order was issued for his arrest but he fled with his family to Comitan in the Mexican state of Chiapas. Burkitt knew Kanter and was able to obtain some thirty-five photographs of Kanter's extraordinary archaeological collection. Burkitt's association with Kanter indicates that Burkitt was collecting word lists prior to 1915 (Seler 2003:16).

2. Burkitt records the beginning dates for the indigenous calendar in 1900/1901 at various communities as Chamula and Ixtapa (March 22), San Cristóbal (April 11), San Bartolomé (May 1), Comitan (May 21), Ixtatan (June 10), Coatan (June 30), Solomá and Santa Eulalia (July 20), Jacaltenango (August 9), Chajul (August 29), Nebaj (September 18), Ixtahuacán (October 8), Todos Santos (October 28), Chiantla (November 17), Aguacatan (December 7), Quezaltenango (December 27), Nahuala (January 16), Momostenango (February 5), Rabinal (February 25), and Uspantan (March 17).

3. Burkitt's notes for his informants at Nebaj state: "A 191/200. James Brito. Diego Brito, son of the Indian alcalde of the moment. A young chap who had been a good while at school in Santa Cruz del Quiché, and can read and write. He admits he has forgotten some of his Nebaj talk. He had a tendency to use K'iche' words, or pronunciations. Other Nebaj Indians said that this Brito's talk was corrupted, and I came to think that my notes from him were very untrustworthy, and started over once again with another man. Thomas Reymundo. Cannot read or write, but speaks Spanish middling well. A certain Jacinto, with a companion. Middle-aged men. James (Diego) Guzman, an old man, along with a Jacinto Pérez. I believe the same as the Jacinto of C. E 403. A servant at the church (a couple of words)."

APPENDIX 2: NOTES ON THE CORRELATION OF MAYAN AND GREGORIAN CALENDARS

1. Thompson (1950:123) introduced the term "vague year" to refer to an arithmetic solar year of exactly 365 days as opposed to the astronomic solar year.

APPENDIX 3: AGRICULTURAL CYCLE AND THE K'ICHE'AN CALENDAR

1. Information on the various forms of animal life that damage maize planting in all stages of growth can be found in Stadelman (1940) and Wilson (1972). For the sake of convenience these creatures may be divided into four groups: birds, mammals, reptiles, and insects. The common birds are the boat-tailed grackle (*Cassidix mexicanus*), orioles (*Icterusa* spp.),

bushy-crested jay, various species of sparrows (*Melozone* spp.), and other seed-eating birds in general. Most of them confine their attacks to the newly planted seed but the parrot will also eat the roasting ears and the dried ears. The melodious blackbird (*Dives dives; Euphagus cvanocephalus*) pulls up sprouts and pecks at ripe ears, and the bushy-crested jay (*Cassilopha melanocyanea*) pecks out ear initials and strips husks from ripe ears. The ocellated quail (*Cyrtonyx ocellatus*) digs up new sprouts and seeds, and the oriole pecks at ripe ears. The use of traps, scarecrows, poisoned seed, blowguns, and noise making in the fields either with or without firearms are the means used to combat the birds (Stadelman 1940:121; Wilson 1972). Mammals include a varied number of animals, such as the skunk (possibly *Spilogale, Mephitis*, or *Conepatus* spp.), fox (possibly *Urocyon cinereoargentus guatemalae*), coyote (*Canis latrans*), and dog (*Canis familiaris*). Squirrels (*Sciurus* spp.), raccoons (*Procyon lotor shufeldti*), raccoon pizotes (*Nasua narica* L.), and rats and mice of many species (*Peromyscus* spp.; *Rattus* spp.; *Reithrodontomys* spp.; *Sygmodon* spp.), will remove the seed from the ground. The young shoots are often eaten by rabbits (probably *Sylvilagus floridanus*), deer (*Odocoileus virginianus nelsoni* Merriam), forest pig or collared peccary (*Pecari tajacu nelsoni* Goldman), and white-lipped peccary (possibly *Tayassu pecari* Fisher). The greatest damage is done in the roasting-ear stage, for only the rabbit will not eat roasting ears and this only because of its inability to climb or pull down the stalk. Damage from animals is not very important in the highlands; the deer is destructive, but more so in the bean patches than in the maize field. Traps are often set for the smaller animals and sometimes the seed in poisoned. The dogs of the owner are often tied at the edges of the field to frighten off animals. If they are not tied, they too will eat the roasting ears. The only reptiles that attack maize are lizards (*Sceloporus* spp.), which dig out the seed. Lizards are controlled by the burning of the cut bush or cornstalks, which destroys hiding places. There are numerous kinds of insects that attack maize. The most destructive insect is the gallina ciega (*Melolontha* spp.), a white grub or larva of a beetle. This larva lives in the ground and attacks the roots of maize plant, usually when it is about to tassel. In certain years the damage is very great. Stadelman (1940:122) reports that informants state that this pest is most destructive in the drier years and that it died out after heavy rains. Other common milpa pests include earwig (possibly *Dermoptera* spp.) and moth (palomilla), a main pest of stored ears and shelled grain; the corn weevil (*Sitophilus oryzae* L.) attacks dry maize in husk in the field or in storage; milpa worm (possibly *Dermaptera, Forficulidae, Doru*, or *Lepidoptera* spp.) bores down through new leaves; and the leaf-cutter ant (*Atta* spp.) attacks leaves of maize (Wilson 1972:106). Among worms, the black worm (*Noctuidae* spp.) is an active pest of beans and garden vegetables, attacking the roots, flower, and roasting ear. Other worms include a variety of small root worms, earworms, stalk borers, and measuring worms, which attack roots and flowers of the maize. The grasshopper or locust (*Ampyltropidia* spp.) is a dreaded menace that can completely destroy a maize field (Stadelman 1940:12; Wilson 1972:436). The most common disease of maize plants and ears is smut and *argeño*. Smut (rutzam çi) attacks the ear in an early stage of its development, and *argeño*, a wilt or root rot, attacks the roots of the plant just before or just after the ears have appeared.

REFERENCES

Ajpacaja Tum, Pedro Florentino, Manuel Isidro Chox Tum, Francisco Lucas Tepaz Raxuleu, and Diego Adrian Guarchaj Ajtzalam. 1996. *Diccionario k'iche'*. Antigua Guatemala: Proyecto Lingüístico Francisco Marroquin.

Akkeren, Ruud van. 2000. *Place of the Lord's Daughter: Rab'inal, Its History, Its Dance-Drama*. Leiden: Research School, School of Asian, African, and Amerindian Studies, Universiteit Leiden.

Andrade, Manuel J., Sol Tax, Asael T. Hansen, and Alfonso Villa Rojas. 1938. Ethnological, Sociological, and Linguistic Research. *Carnegie Institution of Washington, Year Book* 37:162–164.

Annals of the Cakchiquels. 1953. *The Annals of the Cakchiquels; Title of the Lords of Totonicapán*. Adrian Recinos and Delia Goetz, trans. Norman: University of Oklahoma Press.

Barrera, Francisco. 1745. Abecedario en la lengua que dizen qiche hecho por . . . Barrera que han solamente se conpone de el modo y realidad con que los Judios la ablan el qual no tiene los significados que tienen los latinos pero aprendiendo en el lai pronunciaciones lesera muy facil el saberla muy bien y con toda brebedad, y este dicho Abecedario lo hize segun lo que alcanzo a peticion de MR. Fr. Gabriel Guerrero. Fr. Alberto Míguez. Manuscript, Princeton University Library.

Basseta, Domingo de. 1698. Vocabulario de la lengua quiche, acompañado de otro quiche-castellano, en el dialecto de Rabinal, con una breve gramática del mismo y un vocabulario castellano-quiché. Manuscript, Bibliothèque Nationale, Paris.

Bowditch, Charles P. 1901. Memoranda on the Maya Calendars Used in the Books of Chilam Balam. *American Anthropologist* 3:129–138.

Brack-Bernsen, Lis. 1977. *Die Basler Mayatafeln: Astronomische Deutung der Inschriften auf den Türstürzen 2 und 3 aus Tempel IV in Tikal.* Basel: Verhandlungen der Naturforschenden Gesellschaft, 86(1–2).

Brasseur de Bourbourg, Charles E. 1857. *Bibliotheque mexico-guatémalienne, précédée d'un coup d'œi sur les études américaines dans leurs rapports avec les études classiques et suivie du tableau par ordre alphabétique des ouvrages de linguistique américaine contenus dans le meme volume; rédigée et mise en ordre d'après les documents de sa collection américaine par M. Brasseur de Bourbourg.* Paris: Maisonneuve.

Brinton, Daniel G. 1893. The Native Calendar of Central America and Mexico: A Study in Linguistics and Symbolism. *Proceedings of the American Philosophical Society* 31: 325–328.

Bunting, Ethel Jane. 1932. Ixtlavacan Quiché Calendar of 1854. *Maya Society Quarterly* 1:72–75.

Bunzel, Ruth. 1952. *Chichicastenango: A Guatemalan Village.* Publication 22. Locust Valley: American Ethnological Society.

Burkitt, Robert. 1930–1931. The Calendar of Soloma and of Other Indian Towns. *Man* 30–31:103–107, 146–150.

Burkitt, Robert J. N.d. Miscellaneous Linguistics Field Notes. In possession of Dr. Elin Danien, University of Pennsylvania Museum, Philadelphia.

Calendario de los indios de Guatemala, 1685, Cakchiquel [copiado en la Ciudad de Guatemala, marzo 1878]. Manuscript, University of Pennsylvania Museum of Archaeology and Anthropology Library, Philadelphia.

Calendario de los indios de Guatemala, 1722, Kiche [copiado en la Ciudad de Guatemala, abril 1877]. Manuscript, University of Pennsylvania Museum of Archaeology and Anthropology Library, Philadelphia.

Cancian, Frank. 1965. *Economics and Prestige in a Maya Community: The Religious Cargo System in Zinacantan.* Stanford, CA: Stanford University Press.

Carmack, Robert M. 1973. *Quichean Civilization: The Ethnohistoric, Ethnographic, and Archaeological Sources.* Berkeley: University of California Press.

Carter, William E. 1969. *New Lands and Old Traditions: Kekchi Cultivators in the Guatemalan Lowlands.* University of Florida, Latin American Monographs, 6. Gainesville.

Colby, Benjamin N., and Lore M. Colby. 1981. *The Daykeeper: The Life and Discourse of an Ixil Diviner.* Cambridge, MA: Harvard University Press.

Cortés y Larraz, Pedro. 1958. *Descripción geográfico moral de la diócesis de Goathemala.* Bibliotheca Goathemala, 20. Guatemala: Sociedad de Geografía e Historia de Guatemala. 2 v.

Coto, Thomas de. 1983. [*Thesavrvs verborv*]: *vocabulario de la lengua cakchiquel v[el] guatemalteca, nueuamente hecho y recopilado con summo estudio, trauajo y erudición*. René Acuña, ed. Mexico: Universidad Nacional Autonoma de Mexico.

Danien, Elin C. 1985. Send Me Mr. Burkitt, Some Whiskey and Wine: Early Archaeology in Central America. *Expedition* 27(3):26–32.

Dürr, Michael. 1987. *Morphologie, Syntax und Textstrukturen des (Maya-) K'iche' des Popol Vuh: Linguistische Beschreibung eines kolonialzeitlichen Dokuments aus dem Hochland von Guatemala*. Reihe Alt-Amerikanistik, Band 2. Bonn.

Edmonson, Munro S. 1965. *Quiché-English Dictionary*. Middle American Research Institute, Tulane University, Publication 30. New Orleans.

Edmonson, Munro S. 1988. *Book of the Year: Middle American Calendrical Systems*. Salt Lake City: University of Utah Press.

Edmonson, Munro S. 1997. *Quiche Dramas and Divinatory Calendars: Zaqi Q'oxol and Cortés: The Conquest of Mexico in Quiché and Spanish; The Bull Dance; The Count of the Cycle and the Numbers of the Days*. Middle American Research Institute, Tulane University, Publication 66. New Orleans.

Falla, Ricardo. 1975. La conversión religiosa: Estudio sobre un movimiento rebelde a las creencias tradicionales en San Antonio Ilotenango, Quiche, Guatemala (1948–70). Doctoral dissertation, Department of Anthropology, University of Texas, Austin.

Fuls, Andres. 2004. Das Rätsel des Mayakalenders. *Spektrum der Wissenschaft* 1:52–59.

García Elgueta, Manuel. 1962. Descripción geográfica del Departamento de Totonicapán. *Guatemala indígena* 8:115–202.

Gates, William E. 1932a. A Lanquin Kekchi calendar. *Maya Society Quarterly* 1(1):30–32.

Gates, William E. 1932b. Pokonchi calendar. *Maya Society Quarterly* 1(2):75–77.

Goubaud Carrera, Antonio. 1935. El Guajxaquip Batz: Ceremonia calendarica indigena. *Anales de la Sociedad de Geografia e Historia de Guatemala* 12:39–52.

Gubler, Ruth, and David Bolles. 2000. *The Book of Chilam Balam of Na: Facsimile, Translation, and Edited Text*. Lancaster, CA: Labyrinthos.

Hernández Spina, Vicente. 1854. Calendario conservado hasta el dia por los sacerdotes del sol en Ixtlahuacan, pueblo descendiente de la nación kiché. Manuscript, University of Pennsylvania Museum of Archaeology and Anthropology Library, Philadelphia.

Hinz, Eike. 1991. *Misstrauen führt zum Tod: Die psychotherapeutischen Beratungsgespräche eines Ratgebers der Kanjobal-Maya*. Hamburg: Wayasbah. 2 v.

Karttunen, Frances E. 1983. *An Analytical Dictionary of Nahuatl*. Austin: University of Texas Press.

Kelley, David. 1976. *Deciphering the Maya Script*. Austin: University of Texas Press.

La Farge, Oliver. 1930. The Ceremonial Year at Jacaltenango. *Proceedings of the International Congress of Americanists* (23 Session, New York, 1928), pp. 656–660. New York.

La Farge, Oliver. 1934. The Post-Columbian Dates and the Mayan Correlation Problem. *Maya Research* 1:109–124.

La Farge, Oliver. 1947. *Santa Eulalia: The Religion of a Cuchumatan Indian Town*. Chicago: University of Chicago Press.

La Farge, Oliver, and Douglas S. Byers. 1931. *The Year Bearer's People*. Middle American Research Institute, Tulane University, Publication 3. New Orleans.

Lehmann, Walter. 1910. Der Kalender der Quiche Indianer Guatemalas. *Anthropos* 6:403–410.

León Chic, Eduardo. 1999. *El corazón de la sabiduría del pueblo maya; uk'u'xal ranima' ri qano'jibal.* Guatemala: Fundación Centro de Documentación e Investigación Maya.

Lincoln, Jackson S. 1942. The Maya Calendar of the Ixil of Guatemala. *Carnegie Institution of Washington, Contributions to American Anthropology and History* 7:97–128. Washington, DC.

Lincoln, Jackson S. 1945. *An Ethnological Study of the Ixil Indians of the Guatemala Highlands.* Microfilm Collection of Manuscripts on Middle American Cultural Anthropology, 1. Chicago: University of Chicago Libraries.

Lothrop, Samuel K. 1929. Further Notes on Indian Ceremonies in Guatemala. *Museum of the American Indian/Heye Foundation, Indians Notes* 6:1–25.

Lothrop, Samuel K. 1930. A Modern Survival of the Ancient Maya Calendar. *Proceedings of the International Congress of Americanists* (23 Session, New York, 1928), pp. 652–655. New York.

Lounsbury, Floyd G. 1978. Maya Numeration, Computation, and Calendrical Astronomy. In *Dictionary of Scientific Biography.* C. C. Gillespie, ed. v. 15, Supplement 1, pp. 759–818. New York: Charles Scribner.

Lounsbury, Floyd G. 1992. A Derivation of the Mayan to Julian Calendar Correlation from the Dresden Codex Venus Chronology. In *The Sky in Mayan Literature.* Anthony F. Aveni, ed., pp. 184–206. Oxford, England: Oxford University Press.

McBryde, F. Webster. 1947. *Cultural and Historical Geography of Southwest Guatemala.* Smithsonian Institution, Institute of Social Anthropology, Publication 4. Washington, DC.

Miles, Suzanna W. 1952. An Analysis of Modern Middle American Calendars: A Study in Conservation. In *Acculturation in the Americas: Proceedings and Selected Papers of the XXIX International Congress of Americanists.* Sol Tax, ed., pp. 273–284. Chicago: University of Chicago Press.

Miles, Suzanna W. 1957. The Sixteenth-century Pokom-Maya: A Documentary Analysis of Social Structure and Archaeological Setting. *Transactions of the American Philosophical Society* 47:733–781.

Miles, Suzanna W. 1965. Summary of the Preconquest Ethnology of the Guatemala-Chiapas Highlands and Pacific Slopes. In *Archaeology of Southern Mesoamerica.* Gordon R. Willey, ed., pp. 276–287. Handbook of Middle American Indians, 2. Austin: University of Texas Press.

Morán, Pedro. 1720. Bocabulario de solo los nombres de la lengua pocomam. Manuscript, Bibliothèque Nationale, Paris.

Popol Vuh. 1950. *Popol Vuh: The Sacred Book of the Ancient Quiché Maya.* Delia Goetz, Sylvanus G. Morley, and Adrian Recinos, eds. Norman: University of Oklahoma Press.

Rodriguez, Raquel, and Mario Crespo M. 1957. Calendario cakchiquel de los indios de Guatemala (1685). *Antropología e historia de Guatemala* 9:17–27.

Rosales, Juan de Dios. 1949a. *Notes on Aguacatan.* Microfilm Collection of Manuscripts on Middle American Cultural Anthropology, 24. Chicago: University of Chicago Libraries.

Rosales, Juan de Dios. 1949b. *Notes on San Pedro La Laguna*. Microfilm Collection of Manuscripts on Middle American Cultural Anthropology, 25. Chicago: University of Chicago Libraries.

Sáenz de Santa María, Carmelo. 1940. *Diccionario cakchiquel-español*. Guatemala: Sociedad de Geografía e Historia de Guatemala.

Sapper, Karl T. 1925. Uber Brujeria in Guatemala. *Proceedings of the International Congress of Americanists* (21 session, Goteborg), pp. 391–405. Göteborg: Göteborg Museum.

Satterthwaite, Linton, and Elizabeth K. Ralph. 1960. New Radiocarbon Dates and the Maya Correlation Problem. *American Antiquity* 26:165–184.

Schalley, Andrea C. 2000. *Das mathematische Weltbild der Maya*. Frankfurt am Main: Peter Lang.

Scherzer, Karl. 1855. *Sprache der Indianer Central-Amerika's; während seinen Mehrjährigen Reisen in den Verschiedenen Staaten Mittel-Amerika's*. Wien: K. K. Hof und Staatsdruckerie.

Scherzer, Karl. 1856. *Die Indianer von Santa Catalina Istlavacan; Ein Beitrag zur Cultur-Geschichte der Urbewohner Central-Amerikas*. Wien: K. K. Hof und Staatsdruckerie.

Scherzer, Karl. 1864. *Aus dem nature- und Völkerleben im Tropischen America*. Leipzig: G. Wigand.

Scherzer, Karl. 1954. Los indios de Santa Catarina Istlavacan (pie de mujer): Una contribución para la historia de la cultura de los habitantes originales de Centro America. *Antropología e historia de Guatemala* 6:13–21.

Scholes, France V., Ralph L. Roys, Eleanor B. Adams, and Robert S. Chamberlain. 1946. History of Yucatan. *Carnegie Institution of Washington, Year Book* 45:217–221.

Schuller, Rudolf R. N.d. Cholbal k'ih and Ahilabal k'ih: Anonymous Quiche Manuscript. Rudolph Schuller, trans.; Oliver La Farge and J. Alden Mason, eds. Manuscript, University of Pennsylvania Museum Library. Philadelphia.

Schultze Jena, Leonhard. 1933. *Leben, Glaube und Sprache der Quiche von Guatemala*. Indiana, 1. Jena: Gustav Fischer.

Schultze Jena, Leonhard. 1946. *La vida y las creencias de los indígenas quiches de Guatemala*. Guatemala: Sociedad de Geografia e Historia de Guatemala.

Schultze Jena, Leonhard. 1947. *La vida y las creencias de los indígenas quiches de Guatemala*. Biblioteca de Cultura Popular, 49. Guatemala.

Schulz Friedemann, Ramon. 1955. Nueva sincronología maya. *El México Antiguo* 8:225–232.

Seler, Eduard. 2003. *The Ancient Settlements of Chacula, in the Nentón District of the Department of Huehuetenango, Republic of Guatemala*, ed. John M. Weeks. Lancaster, CA: Labyrinthos.

Solano, Félix. 1580. Vocabulario en la lengua castellana y cakchiquel, que se llama cak-chiquel-chi. Manuscript, Princeton University Library, Princeton, NJ.

Spinden, Herbert J. 1924. *Reduction of Mayan Dates*. Peabody Museum of American Archaeology and Ethnology, Harvard University, Papers, 6(4). Cambridge, MA.

Stadelman, Raymond. 1940. Maize Cultivation in Northwestern Guatemala. *Carnegie Institution of Washington, Contributions to American Anthropology and History* 6(33):83–263. Washington, DC.

Steinberg, Michael K., and Matthew Taylor. 2002. The Impact of Cultural Change and Political Turmoil on Maize Culture and Diversity in Highland Guatemala. *Mountain Research and Development* 22(4):344–351.

Suchtelen, Bertho C. van. 1958. Some More Supplementary Notes on the MS Landa and the Leyden Plate. Oegstgeest. Manuscript. University of Pennsylvania Museum Library.

Tax, Sol. 1947a. *Miscellaneous Notes on Guatemala*. Microfilm Collection of Manuscripts on Middle American Cultural Anthropology, 18. Chicago: University of Chicago Libraries.

Tax, Sol. 1947b. *Notes on Santo Tomas Chichicastenango*. Microfilm Collection of Manuscripts on Middle American Cultural Anthropology, 16. Chicago: University of Chicago Libraries.

Tax, Sol. 1953. *Penny Capitalism: A Guatemalan Indian Economy*. Smithsonian Institution, Institute of Social Anthropology, Publication 16. Washington, DC.

Tedlock, Barbara. 1982. *Time and the Highland Maya*. Albuquerque: University of New Mexico Press.

Tedlock, Barbara. 1992. *Time and the Highland Maya*, rev. ed. Albuquerque: University of New Mexico Press.

Teeple, John E. 1931. Maya Astronomy. *Carnegie Institution of Washington, Contributions to American Archaeology* 1:29–115.

Termer, Franz. 1930. Zur Ethnologie und Ethnographie des nördlichen Mittelamerika. *Ibero-Amerikanisches Archiv* 4:303–492.

Thompson, J. Eric S. 1932. A Maya Calendar from the Alta Vera Paz, Guatemala. *American Anthropologist* 34(3):449–454.

Thompson, J. Eric S. 1934. Maya Chronology: The Fifteen Tun Glyph. *Carnegie Institution of Washington, Contributions to American Archaeology* 2(11):243–254.

Thompson, J. Eric S. 1950. *Maya Hieroglyphic Writing: Introduction*. Publication 589. Washington, DC: Carnegie Institution of Washington.

Thompson, J. Eric S. 1971. Preface. In *Maya Hieroglyphic Writing: An Introduction*, 3rd ed. Norman: University of Oklahoma Press.

Thompson, J. Eric S. 1972. *A Commentary on the Dresden Codex: A Maya Hieroglyphic Book*. Philadelphia: American Philosophical Society.

Title of the Lords of Totonicapan. 1953. *Title of the Lords of Totonicapan*. Dionisio José Chonay and Delia Goetz, eds. Norman: University of Oklahoma Press.

Tovilla, Martin A. 1960. *Relacion histórico-descriptiva de las provincias de la Verapaz, el Manché y Lacandón en Guatemala* (1635). Guatemala: Editorial Universitaria.

Varea, Francisco de. ca. 1635. Calepino en lengua cakchiquel, por fray Francisco de Varea, hijo de esta S. Provincia del S. S. Nombre de Jesús de religiosos de N.P.S. Francisco de Guatemala. Manuscript, American Philosophical Society, Philadelphia.

Vico, Domingo de. 1550. Vocabulario cakchiquel con advertencia de los vocablos de las lenguas k'iche' y tzutuhil. Manuscript, Bibliothèque Nationale, Paris.

Vogt, Evon Z. 1969. *Zinacantan: A Maya Community in the Highlands of Chiapas*. Cambridge, MA: Harvard University Press.

Wagley, Charles. 1941. *Economics of a Guatemalan Village*. American Anthropological Association, Memoir 58, Washington, DC.

Willson, Robert W. 1924. *Astronomical Notes on the Maya Codices*. Peabody Museum of American Archaeology and Ethnology, Harvard University, Papers, 6(3). Cambridge, MA.

Wilson, Michael R. 1972. A Highland Maya People and Their Habitat: The Natural History, Demography and Economy of the K'ekchi'. Doctoral dissertation, Department of Geography, University of Oregon, Eugene.

Ximénez, Francisco. 1929–1931. *Historia de la Provincia de San Vicente de Chiapa y Guatemala de la Orden de Predicadores, compuesta por el R.P.Pred.Gen. fray Francisco Ximénez, hijo de la misma provincia de orden de N.Rmo.P.M.G. Fr. Antonio Cloché*. Bibliotheca Goathemala, 1–3. Guatemala: Sociedad de Geografía e Historia.

Ximénez, Francisco. 1985. *Primera parte del Tesoro de las lenguas cakchiquel, quiché, y tzutuhil, en que las dichas lenguas se traducen en la nuestra, española,, de acuerdo con lops manuscritos redactados en la Antigua Guatemala a principios del siglo XVIII, y conservados en Córdova (España) y Berkeley (California)*. Carmelo Sáenz de Santa María, ed. Publicación Especial, 30. Guatemala: Academia de Geografía e Historia de Guatemala.

INDEX

218